The
Fundamentals
of
Imaging

From Particles to Galaxies

The
Fundamentals
of
Imaging

From Particles to Galaxies

Michael M. Woolfson
University of York, UK

Imperial College Press

Published by

Imperial College Press
57 Shelton Street
Covent Garden
London WC2H 9HE

Distributed by

World Scientific Publishing Co. Pte. Ltd.
5 Toh Tuck Link, Singapore 596224
USA office: 27 Warren Street, Suite 401-402, Hackensack, NJ 07601
UK office: 57 Shelton Street, Covent Garden, London WC2H 9HE

British Library Cataloguing-in-Publication Data
A catalogue record for this book is available from the British Library.

THE FUNDAMENTALS OF IMAGING
From Particles to Galaxies

45.60

ISBN-13 978-1-84816-684-4
ISBN-10 1-84816-684-2
ISBN-13 978-1-84816-685-1 (pbk)
ISBN-10 1-84816-685-0 (pbk)

Typeset by Stallion Press
Email: enquiries@stallionpress.com

Printed in Singapore by Mainland Press Pte Ltd.

Contents

Preface

For highly developed living species, especially mankind, the creation of images is an important factor in their awareness of being alive and of consciousness — the understanding of one's identity and place in the world. Image formation is also an essential process in the advance of man's knowledge of the world and the Universe in which he lives. It is through images that we comprehend and interpret our surroundings and react to what happens in it. Aboriginal rock paintings in caves in Australia have been dated to more than 50,000 years ago and they are a testament to the importance that images have always had for mankind, even when living in primitive societies. When we sleep, and are in a state without consciousness, we dream and create images, often bizarre and unrealistic, that sometimes, but not always, can be recalled when we wake. In fairly recent times the interpretation of dreams was regarded by the psychoanalyst Sigmund Freud as an important tool in the treatment of psychiatric illness. Somewhat further back, the biblical story of Joseph (he of the multicoloured garment) hinged on his interpretation of the Egyptian Pharaoh's dreams in terms of an agricultural forecast for the years ahead.

The crudest type of imaging is just the determination that some object exists at a particular place without being able to discern its form in any way. This is usually referred to as *detection* rather than imaging and although it is a rather poor relation of detailed imaging it is of importance in many spheres of human activity — for example, in air traffic control or submarine detection. At the other end of the scale there are cases where the nature of an object is revealed by imaging in the very finest detail.

Of all the senses sight is king. There are the senses that require contact; we can taste what is in our mouths and we can touch what we can reach. There is a party game in which an object is placed in a bag and just by feel alone one must try to identify what it is. Most people are quite successful at this for a simple object and can create a mental image of it. They are less successful for objects with a complicated structure — although blind individuals, whose sense of touch is far more sensitive and well developed, may do better. If the object is patterned in some way, e.g. it is a photograph, then no degree of sensitivity of touch could determine the full information that it offers to those viewing it directly.

Smell is a sense that has a limited range for humans but is more developed in some other species for which it has survival value, giving notice of the proximity of either prey or predator. Hearing for humans has a much larger range than smell and very loud noises, such as explosions due to major accidents or to artillery fire, can be heard over hundreds of kilometres. However, for sight the range is, to all intents and purposes, infinite. We may not be able to see the details and the exact form of a distant star or galaxy with the naked eye but if the intensity of light is sufficiently high then we can detect that it is there.

The first chapters of this book deal with the structure of the human eye, how it produces images and how it may have evolved. It cannot be guaranteed that the image of an object perceived by a viewer precisely represents its true form. If it is seen with the assistance of an optical system then there may be distortions introduced by that system; the eye itself is an optical system that can introduce distortion in a true image and, when this is troublesome, spectacles are used to reduce the distortion to an acceptable level. In other cases in which distorted or corrupted images have been formed, scientists have devised processes, referred to as image processing, mathematically based and implemented by computers, by which a better image can be obtained.

What we have been describing so far is either the everyday detailed viewing of objects, ranging from the finest print to a landscape or a distant cloud, or just the detection of a luminous object, such as a star. There are contexts, usually of a scientific nature, where it is required to observe objects so small that they cannot be resolved by an unaided eye or, if very large, e.g. a galaxy, so distant that they can only be seen as point light sources, or even not at all, by the unaided eye. In such cases there are optical devices, microscopes and telescopes that enable a more-or-less true image of the object to be seen and recorded.

The implication in the above discussion is that the objects are being viewed, with or without instrumental aids, with visible light. However, there are many radiating objects which emit too little visible light to be seen directly but do emit other kinds of radiation that can reveal an image of their form and structure by the use of suitable instruments. In the field of astronomy the range of electromagnetic radiation used for detecting or imaging extends from radio waves with wavelengths a few tens of centimetres to γ-radiation with wavelengths of order 10^{-14} metres.

There is an important difference between seeing this book and seeing a candle flame. A candle flame, like a star, itself emits visible light so that there is no need for any other source of radiation to be used in order for it to be seen. However, a book emits no visible radiation so, to see this book, it is necessary first to illuminate it and then to recombine the radiation scattered from it to form an image. There is a general rule that when producing an image the resolution attainable is, at best, no better than the wavelength of the wave motion being used. If the object to be seen is very tiny and fine details of it are to be imaged then normal visible light may not give the resolution required and radiation of a shorter wavelength, e.g. ultraviolet light, may then be used.

There are many important categories of image formation that involve the recombination of a wave motion scattered by an object. Notice that I have used the words 'wave motion' here because the scattering need not be of light waves, or some other form of electromagnetic radiation, but could be of another kind. Images can be formed by the use of ultrasound — sound waves of very high frequency — and this has both medical and engineering applications. Indeed, it is by the use of ultrasound that bats, the flying mouse-like creatures, detect their prey, flying insects, at night. Electrons that, according to modern theory and as confirmed experimentally, can have wave-like properties give another possible wave motion for imaging.

One of the great scientific achievements of the twentieth century is the development of methods of detecting the positions of atoms in crystals by the technique of x-ray crystallography — creating an image of a crystal structure. This has enabled important developments in many branches of science, no more so than in biochemistry and medicine for which knowledge of the crystal structures of proteins, of deoxyribonucleic acid (DNA) and other macromolecules has enabled enormous advances to be made. However, in this case there are no physical ways of combining the x-rays scattered by the crystal to form an image by any physical process; instead the pattern of scattered radiation must be handled mathematically and the image is the result of calculation. The great advances made by x-ray crystallography would have been impossible without the availability of powerful computers.

In this book we shall explore a wide range of techniques for detecting and imaging objects of various kinds, ranging in size from galaxies down to the fundamental sub-atomic particles of matter. There will also be brief descriptions and explanations of the objects being imaged; for example, before ways of imaging exotic particles are described in the final chapter there will be an explanation of what those particles are and how they relate to the physical world. We shall be describing how these various techniques can both be of benefit to mankind, for example, in medical imaging and by making flying in aircraft a safer activity, and also extend our knowledge of the Universe and all its contents. It will not be a formal, highly mathematical and comprehensive treatment but one that reveals the wide range of imaging techniques that are available. It should be accessible to those with an interest in science and who have studied science at school to an intermediate level.

Chapter 1

The Human Visual System

The visual systems of humans and other vertebrates are so complex and ingenious in the way that they operate that they were offered as a reason for doubting the theory of evolution given in 1859 by Charles Darwin (1809–1882). The argument put forward was that it was inconceivable that such a complicated system could arise in a gradual incremental way via mutations and natural selection. The visual system has several components and *all* of those components have to be present for the system to operate effectively. The opponents of Darwin argued that the human visual system can only have been a consequence of intelligent design and Darwin accepted that it was a case that needed to be answered — although he was adamant that evolutionary processes were capable of doing what was needed. In this chapter we shall begin with a description of the human visual system and then, in the following chapter, consider by what processes it could have evolved from a simple origin.

1.1. The Optical System

The human visual system has three distinct components: an optical system that produces an image; a neural network system that converts that image into electrical impulses; and an interpretation system, located in the brain, which converts the electrical signals into the sensation by which the object being imaged is visualized. The optical system is very simple and produces an image in the same way as is done by a simple convex lens. This is illustrated in Fig. 1.1 where the image of an arrow is projected onto a screen.

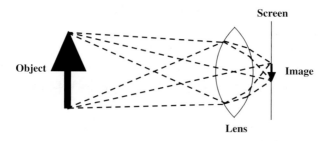

Figure 1.1   Producing an image on a screen with a simple convex lens.

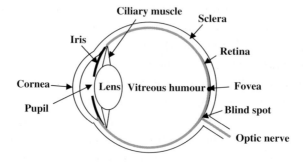

Figure 1.2   A schematic representation of a human eye.

The lens operates in such a way that rays of light, leaving a point on the object, travelling in different directions but all passing through the lens, are brought together at a point on the screen. From this diagram we can understand the conditions for producing a sharp image. If the screen were moved either forwards or backwards from the position shown then the light from a point on the object would be spread out into a circular patch and the image would be fuzzy. The same effect would come about as the result of keeping the position of the lens and screen constant but using a lens of a different focal length that bent the light differently.

Figure 1.2 shows a very simplified diagram of the structure of the human eye, and we now relate its operation in producing a sharp image to that of the simple lens. The combination of the *cornea* and *lens* produces an image on the *retina*, which is the equivalent of the screen in Fig. 1.1. However, for the eye, the distance of objects for which sharp images must be produced is variable, from very close when reading a book to very far when viewing a distant landscape. In the case of a lens and screen, if the distance of the object is changed then the distance of the screen from the lens can be changed to produce a sharp image. For the eye, the lens-to-screen distance is fixed so the only way to produce a sharp image is to change the focal length of the lens. This is done by means of the *ciliary muscles* that, by squeezing the lens, can change its shape and hence its focal length. The contribution of the cornea in the focusing process does not change but in a normal eye the total possible variation in the focal length of the lens is sufficient to allow clear images of both near and far objects to be produced. This variation is called *accommodation* and for some individuals it is unable to cover the full range of distances. Those who can see only close objects in focus suffer from short-sightedness, or *myopia*, while those able to see only far objects in focus suffer from far-sightedness, or *hyperopia*. In both cases spectacles can usually deal with the problem. Another common problem is that of *astigmatism* in which the curvature of the cornea is different in different directions. This means that light rays coming from the object and lying in different planes cannot all be brought into focus simultaneously. It is a common fault but normally so slight that it causes no difficulties to an individual. Where it is more severe then, once again, suitable spectacles can compensate for it.

The bulk of the eye is filled with a transparent gel-like material known as the *vitreous humour*. The *sclera*, seen in Fig. 1.2, together with the cornea, forms its outer container. While the cornea is transparent, as it must be to transmit light, the sclera is a milky-white colour — indeed, it is what is commonly called the *white of the eye*. The eye functions most efficiently at moderate light levels and the *iris*, the coloured disk that gives eye colour, controls a variable circular aperture that, by opening and closing, adjusts the amount of light falling on the retina. This aperture, known as the *pupil*, is seen as a black circle in the centre of the iris. In dim light the pupil opens wide to let through as much light as possible while in very bright conditions it closes up to reduce the amount of light entering the eye.

1.2. The Photoreceptors

Having formed an image on the retina, the information describing it must be transformed into electrical signals that are transmitted to the brain. The first stage of this process is done by photoreceptors distributed all over the retina. There are about 150 million of them, distributed most densely in the centre of the field of view in the region shown as the *fovea* in Fig. 1.2, with the density falling off with distance from the fovea. These photoreceptors are of two kinds. One kind consists of long thin receptors called *rods*, which are very sensitive and can record very low light levels. They give what is known as *scotopic vision*, which would, for example, operate at night with the Moon as illumination. The second kind of receptor is stubbier; they are called *cones* and they are responsible for *photopic vision*, which occurs under everyday normal light conditions.

Apart from their sensitivity, rods and cones have another important difference — in the degree to which they give the perception of colour. There is no perception of colour with scotopic vision — just shades of grey, although it sometimes seems that moonlight has a bluish tinge. By contrast, photopic vision is fully chromatic and this is because there are three kinds of cones with different colour responses, illustrated in Fig. 1.3.

In every part of the retina there is a mixture of the different types of cone and the relative responses of the three types gives the colour perceived. For example, if the red response is somewhat greater than the green response, with virtually no blue response, as indicated by the dashed line in Fig. 1.3, then the colour seen will be yellow. As another example, if the largest contribution is from green with a smaller contribution from red and a significant contribution from blue, as indicated by the dotted line in the figure, then the colour seen will be cyan (green-blue). In practice the light received by the eye normally covers a range of wavelengths but the relative response of the three kinds of receptor will still indicate the colour perceived. If all the receptors are affected equally then the colour seen is white. Usually the relative responses of the three kinds of receptor correspond to that from some wavelength of the spectrum plus equal responses corresponding to white light, in which case the

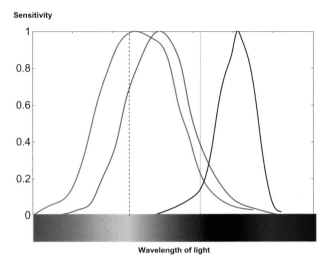

Figure 1.3   The relative sensitivity for each type of cone in the retina, each normalized to a maximum of unity.

colour seen is an unsaturated spectral colour — sometimes known as a *pastel shade*. Finally, if the responses are strong for blue and red but weaker for green then this corresponds to no colour in the spectrum and gives the colour sensation of purple.

The strength of the responses of the three kinds of receptor in terms of their output do not have the same maximum value, as indicated in Fig. 1.3, but the essentials of the process of colour perception are as described.

The foveal region contains only cones, which are tightly packed there, hence giving the greatest acuity of vision at the centre of the field of view. Moving outwards from the fovea, the proportion of rods becomes greater, giving better peripheral vision under low light conditions but with deterioration of acuity as the density of photoreceptors is lower. The responses of individual rods and cones influence the electrical signals that travel along individual optic nerve fibres that pass into the visual cortex in the brain. However, there is not a one-to-one relationship between photoreceptors and optic nerve fibres; the responses of about 150 million receptors are funnelled into about 1 million optic nerve fibres. This funnelling process is carried out by four intermediate layers of specialist cells that mix and process the data from the photoreceptors before they reach the optic nerve fibres. There is no complete understanding of the way this system operates but in order to appreciate the processes that occur we first need to know something about the way that nerve cells work and communicate with each other.

1.3. The Way that Nerve Cells Operate and Communicate

The behaviour of all nervous systems, including that of the eye, is extremely complex and involves physical processes triggered by various chemical agents. Here we give just a very elementary description of the action of a typical nerve cell, sufficient for the purpose of describing the general processes that occur in the neural networks of the eye.

Figure 1.4 shows a schematic nerve cell, or *neuron*. At one end there are *dendrites*, connections that receive information from other neurons. At the other end is the *axon*, a channel along which the electrical nerve impulse travels and the terminus of which communicates with other neurons. The interior of the cell contains an excess of potassium ions, K^+, potassium atoms lacking one electron, which gives it a positive charge. Negatively charged molecules within the cell counterbalance these positive charges to some extent. Outside the cell there is an excess of sodium ions, Na^+, and in the resting state the electric potential outside the cell is greater than that inside the cell. The ions are able to permeate through the membrane that acts as a sheath for the cell but there are chemicals within the membrane that act like ion pumps, a separate one for each ion, so maintaining the relative concentrations of the two kinds of ion. A cell in this situation is said to be *polarized*.

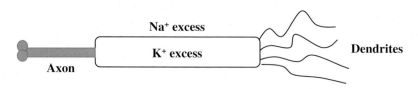

Figure 1.4   A schematic nerve cell.

The connections between axons and dendrites are called *synapses* and for all cells, except those in the brain, the transmission of the signal across a synapse is by chemicals known as *neurotransmitters*. These traverse the tiny synaptic gaps and enter the receptor cell via its dendrites where they become attached to chemical receptors. Synapses can either excite activity or inhibit activity in the receptor cell dependent on what happens when the neurotransmitter arrives. Activity is stimulated when the neurotransmitter opens up sodium channels so that more sodium ions enter the cell. When this happens the potential difference between the inside and outside will reverse — the cell is then *depolarized* and an electrical impulse is transmitted through the neuron. This occurs when the depolarization potential difference reaches a certain critical level — the *threshold potential*. As sodium channels are opened up in one part of the cell so the potential difference reaches a critical level to open up sodium channels further along the cell and this way, in a few milliseconds, the electrical pulse traverses the neuron, moving towards the end of the axon. When the electric pulse arrives at the terminus of the axon, the membrane there allows the passage of doubly charged calcium ions, Ca^{2+}, into the cell and this triggers the release of the neurotransmitter towards the dendrites of the next cell. The form of the action potential, i.e. the potential difference across the membrane, at each point of the neuron as a function of time is shown in Fig. 1.5. As more and more sodium ions enter the cell so this stimulates the potassium channels to open and potassium ions move to the outside of the cell until eventually the cell returns towards its polarized state with a slight overreaction to become *hyperpolarized*, where the difference between the outside and inside potential is higher than the normal resting state of the neuron. When the cell is hyperpolarized the sodium and potassium pumps return to their normal mode of action and restore the condition in which there is an excess of sodium ions on the outside and an excess of potassium ions on the inside. While this state is being restored the neuron cannot be stimulated; it is then in its *refractory period*.

The rate at which action potential spikes are generated depends on the strength of the stimulation but at a certain level of stimulation a stage of saturation is reached and no greater rate of spike

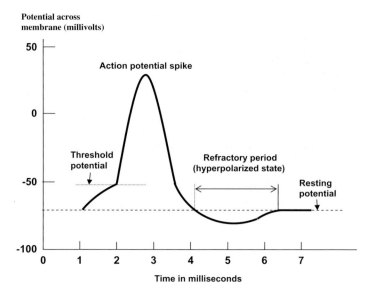

Figure 1.5   The action potential of a neuron. The potential across the membrane is the potential inside minus the potential outside and hence is negative in the resting state.

discharge is then possible, no matter how strong the stimulation. The neuron reaction we have just described is when it is excited by the oncoming neurotransmitter. However, there can also be an inhibitory response of the cell, when it is the potassium rather than the sodium channels that are stimulated to open. In this case potassium ions flood out of the cell, it becomes hyperpolarized, and the impulse is stopped dead in its tracks through the system. Both excitatory and inhibitory reactions are important in the functioning of the optical neural network system.

1.4. The Neural Network of the Eye

There are four layers of neural cells between the photoreceptors and the optic nerve fibres that transmit the information about the image on the retina to the brain. These layers consist of horizontal cells, bipolar cells, amacrine cells and ganglion cells. A representation of these layers of neurons is shown in Fig. 1.6. For some of these

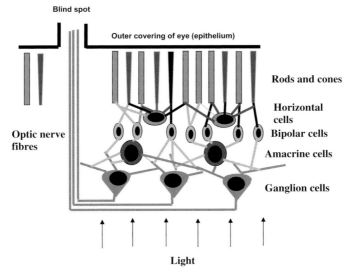

Figure 1.6   A schematic representation of the neural network structure of the eye.

neurons — horizontal cells, bipolar cells and many amacrine cells — the response is not to give a series of potential spikes, the frequency of which indicate the strength of the stimulation, but rather to generate a steady potential that governs the rate of emission of neurotransmitter at the cell axon. In the visual neural network system it is only ganglion cells and some amacrine cells that generate spike potentials that pass into the optic nerve fibres.

The first thing to notice is the apparently anomalous feature that before light reaches the photoreceptors it has to traverse all the other layers of neurons. Apart from the obvious disadvantage that the light will be somewhat absorbed and scattered by the time it reaches the rods and cones, there is also the slight nuisance that light is blocked in the part of the retina where the optic nerve enters the retina, leading to a *blind spot* in the visual field (Fig. 1.2). Detecting this blind spot can be done by a very simple experiment. Draw two dots on a sheet of plain paper about 10 cm apart. Then, with the dots horizontal, just using the right eye, focus onto the left-hand dot. If you start with the paper about 40 cm away and slowly bring it inwards at a distance of about 30 cm the right-hand dot will disappear — its image on the retina is at the position of the blind spot. It is interesting to note that *cephalopods*, creatures of a class including the octopus, squid and cuttlefish, have optical systems of similar quality to that of humans but with the advantage that the layers of neurons are reversed so that light falls on the photoreceptors directly. For this reason there is no blind spot since the axons of the ganglion cells, forming the optic nerve fibres, can link with the brain without traversing the layer of photoreceptors.

We noted at the beginning of this chapter that those opposing Darwin's theory of evolution used the complexity of the eye as an argument for intelligent design. The evolutionary biologist Richard Dawkins (b. 1941) has used the perverse ordering of the neural layers of the human eye to argue *against* intelligent design on the basis that a really intelligent designer would not have incorporated an obvious flaw!

The primary linkage through the neural network is from photoreceptors to bipolar cells and from bipolar cells to ganglion cells. There are several types of bipolar cell, all of which respond to stimulation by varying their internal steady potentials. The first division of types is between those connected to rods and those connected to cones; no bipolar cell has connections to both rods and cones. There is also a further subdivision of bipolar cells into the categories ON and OFF. When a photoreceptor is in the dark it releases a neurotransmitter known as *glutamate* (glutamic acid). Glutamate inhibits the responses of ON bipolar cells and stimulates the responses of the OFF cells. When the photoreceptors are illuminated they generate less glutamate and this has the effect of polarizing the ON bipolar cells, and hence activating them, while hyperpolarizing the OFF cells, so causing them to become inactive. Thus the total magnitude of signal that passes on from the bipolar cell level in the form of neurotransmitter substance is not greatly affected by intensity of the illumination of the photoreceptors; the ON cells pass on more and the OFF cells less. Rather it is the relative response of ON and OFF bipolar cells in any tiny region of the retina that convey the information about the illumination of the connected photoreceptors.

Some of the signals from photoreceptors are sent to horizontal cells. From Fig. 1.6 it is clear that each horizontal cell is receiving information from several photoreceptors and also that horizontal cells have links with bipolar cells. This pooling of information from several photoreceptors is part of the funnelling process by which signals from 150 million photoreceptors eventually lead to spike action potentials in just 1 million optic nerve fibres. However, horizontal cells have another function, that of modifying the response of photoreceptors to light. When the horizontal cells receive glutamate from the photoreceptors they are induced to produce their own neurotransmitter, known as *gaba* (gamma-aminobutyric acid), which is passed back to the photoreceptor and reduces its activity. This leads to a negative feedback mechanism. With more light shining on the photoreceptors there is a decrease in the production of glutamate that, in its turn leads to a reduction in the production of gaba in the horizontal cell. However, since gaba inhibits the activity of the photoreceptors, a reduction in gaba feeding back to the photoreceptor leads to an increase in glutamate production. This feedback mechanism reduces the effect of light on the photocell. Without the feedback mechanism, at a certain level of illumination of the photoreceptors there would be no glutamate production. Any increase of illumination beyond that level would give no change of behaviour of the photoreceptor and this would define the range of intensity that the photoreceptor could discriminate. With the feedback mechanism the reduction in the production of glutamate with increasing illumination is lessened, which increases the range of intensity of illumination that can be discriminated before saturation of the photoreceptor is reached.

Electron micrographs of retinal networks show that rod bipolar cells do not directly link with ganglion cells but only indirectly via amacrine cells. The exact role of amacrine cells is not well understood; clearly they are part of the process by which the signals from the photoreceptors are channelled into many fewer ganglion cells but what other function they might serve is uncertain.

Looking at the neural network as a whole, what we can say is that each ganglion cell is influenced only by the light falling on a particular small patch of the retina. When an image is perceived, the greatest information is gained from boundaries where there is a sharp change of intensity or colour; the visual system has a mechanism for emphasizing such boundaries. Each ganglion cell is influenced in an excitatory fashion by a small circular patch of the retina and in an inhibitory fashion by an annular ring around the circular patch. To see how this gives edge enhancement we consider a one-dimensional illustration shown in Fig. 1.7. The case we take is that the rate of

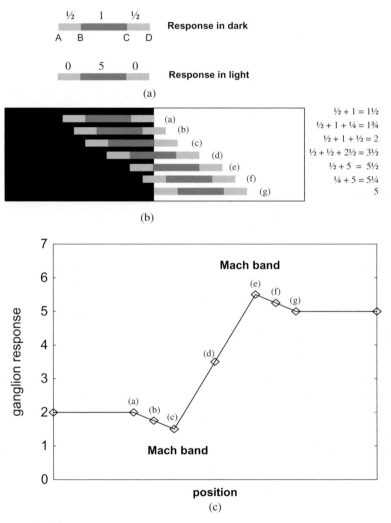

Figure 1.7   (a) The response of the central region in the dark and in uniform illumination. (b) The responses for ganglion cells in different positions in relation to the boundary. (c) The formation of Mach bands.

firing of the ganglion cell, with all its region of influence in the dark, is one unit from the excitatory region, represented in one dimension as BC in Fig. 1.7(a) and one half-unit from each of the inhibitory regions, indicated as AB and CD. However, if the total area is illuminated then the excitatory region gives five units of response while the inhibitory regions give no response — i.e. less than in the dark. Figure 1.7(b) shows the responses of ganglion cells at positions a, b, c, d, e, f and g with the regions of influence shown within a field of view with a sharp delineation between darkness and uniform illumination. A plot of these results, in Fig. 1.7(c), shows that in the field of view, close to the position of the boundary between the dark and illuminated regions, a bump and trough occur. These features, known as *Mach bands*, are a well-known visual phenomenon and help to accentuate the definition of the boundary. Even where there is just a gradation of intensity the formation of Mach bands amplifies the effect and makes the variation more evident. A similar phenomenon, but more complicated to explain, helps the eye to distinguish slight differences of hue. In the more realistic two-dimensional case, with a circular excitatory region surrounded by an inhibitory annulus, the system gives the Mach band effect for any orientation of boundary.

1.5. The Visual Cortex

There is some, although an incomplete, understanding of the processes that occur within the optical neural network. When we come to the problem of how the stream of action potentials moving along the optic nerve fibres is translated into our mental image of the world around us then we must admit that we understand very little. The processing of the information takes place within the *visual cortex*, an area at the rear of the brain. By carrying out experiments in which the electrical activity of various parts of the visual cortex is measured in relation to visual stimuli of different kinds, or in different parts of the visual field, it is possible to identify regions that specialize in certain activities — for example, in detecting motion or in processing signals from the lower half of the visual field. In this way we are able to find out *where* the processing of information occurs, but not *how* it occurs.

One of the most remarkable capabilities of the visual system as a whole, but primarily of the visual cortex, is that of pattern recognition. When we see the letter 'e' in any orientation we instantly recognize it as that letter. Every individual has his own distinctive handwriting, yet the postal service manages to deliver the vast majority of mail to the correct destination.

Chapter 2

The Evolution of the Eye

Living organisms occur in a great range of complexity from the simplest single-celled bacterium to the most intelligent life form, mankind. To determine how even just a bacterium could form from basic chemical constituents is a problem that has not been solved. However, Darwin's theory of evolution tells us that once life exists in its simplest form then its complexity can increase by a series of steps, each very small and almost imperceptible, eventually giving an end product remote from where it started. As an analogy we may take the visible spectrum with wavelengths going from 400 nm at the blue end[1] to 700 nm at the red end. Any two wavelengths separated by one-hundredth of a nanometre are impossible to distinguish by their appearance — yet 30,000 such steps give a change from blue to red.

A characteristic of Darwin's theory is that each step giving permanent change has to confer some benefit to the organism. An argument against the theory — and one that must be addressed — is that there are some features of highly developed organisms that are so complicated that it seems that there can be no possible pathway from something very simple. As an example we can take the case of haemoglobin. This is a large molecule, consisting of four similar units, that is present in blood and has the function of transporting both oxygen and carbon dioxide through the body. Since it is so large, very small local movements throughout the molecule can cause large movements of the surface and there are two distinct configurations

---

[1]1 nanometre (nm) is 10^{-9} m.

of the molecule giving it different behaviour. In one configuration it picks up oxygen from the lungs and deposits carbon dioxide, which is then exhaled. In the other configuration it deposits oxygen at muscle sites, where it is needed for chemical reactions that generate energy, and picks up carbon dioxide, the waste product of those reactions. The argument to be answered by evolutionary theory is that while there are many forms of haemoglobin, varying with the source organism, they are all large molecules because size is essential for its function. The problem is to discover how this molecule could have evolved from something much smaller and simpler that, as far as we can tell, could not carry out the same functions even poorly. While the complexity of the eye is also sometimes quoted as an argument against evolution theory it *is* possible to envisage an evolutionary pathway from simplicity to complexity for the eye, albeit neither in detail nor with certainty.

2.1. Plants and Light

It is a matter of observation that many plants move when exposed to light — for example the leaves of a tree will orient themselves to give maximum exposure to sunlight. This is advantageous to the plant because it is by the action of sunlight on chlorophyll, the green pigment found in leaves, that water, drawn up by the plant's roots, chemically combines with carbon dioxide in the atmosphere to give cellulose, the substance of the plant, and oxygen, which is essential to animate life. Charles Darwin devoted considerable effort in trying to understand this phenomenon of *phototropism*, movement in plants, which he wrote about in his book *The Power of Movement in Plants*.[2] More recent work has shown that one of the main agencies in causing this motion is the chemical indole-3-acetic acid (IAA), one of a class of plant hormones called *auxins*. When light falls on the plant from a particular direction the IAA on the *other* side of the plant produces an elongation of the fibres in its vicinity. This has the effect of bending the plant towards the light (see Fig. 2.1).

---

[2]Darwin, C. (1880) *Power of Movement in Plants*, London, John Murray.

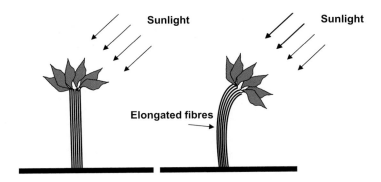

Figure 2.1   The bending of a plant towards light.

If inanimate living forms like plants can be responsive to light in their physiological behaviour then the same should be true even for very simple animate life.

2.2. Different Forms of Eye

An eye is a device for detecting light and transmitting the resultant signal to another part of the organism — the brain for higher life forms — where the information can be translated into a physiological response. When we use the word 'eye' we normally do so in relation to the human eye or to a similar type of eye in mammals. Yet, embraced by the definition given above, there are many types of eye — eight to ten is the range usually quoted — and the human eye is not the one with the greatest effectiveness (in the previous chapter we noted the superior design of the retinal structure of the octopus). Here we shall not exhaustively explore all the different forms of eye but just describe the compound eye as an example of a variant type.

The compound eye is common in many arthropods (insects and crustaceans) and Fig. 2.2 shows an electron micrograph (§4.5) of part of the hemispherical surface of the compound eye of the fruit fly, *Drosophilidae*. Each of the humps is a separate lens and these are packed tightly together to form the continuous surface of the eye.

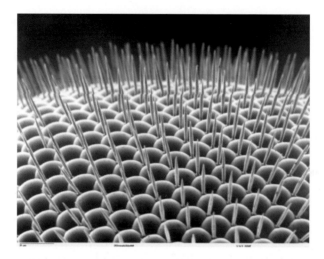

Figure 2.2   The surface of the eye of the fruit fly.

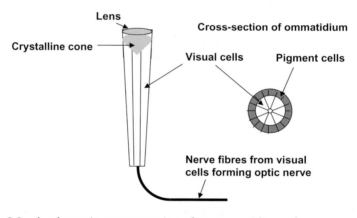

Figure 2.3   A schematic representation of an ommatidium of a compound eye.

The whole eye consists of a very large number of similar units, the *ommatidia*. The structure of a single ommatidium is shown in schematic form in Fig. 2.3.

Each ommatidium is an elongated structure at the head of which is a *lens* followed by a *crystalline cone* that further concentrates the light downwards. The light passes through the cluster of visual cells, at the end of which are the individual nerve fibres forming the optic nerve. Packed round the visual cells are opaque *pigment cells* that prevent light passing between neighbouring ommatidia thus ensuring that each ommatidium detects only the light coming from a small region of the field of view. In this way a halftone picture is built up as a series of small pieces of information from different parts of the field of view. An impression of a very low-resolution image of this kind is shown in Fig. 2.4. Some insects have eyes with very large numbers of ommatidia and hence produce images of reasonable resolution but at best they attain only one or two per cent of the resolution of the human eye and at worst they see the world as rather indistinct patches of light.

In one characteristic, the perception of motion, the compound eye performs extremely well. When an object moves across the field of view individual ommatidia are switched on and off and this is something that is readily detected.

There is more than one form of compound eye and also other types of detection systems that could be described as eyes, albeit they do not form detailed images. However, the vertebrate eye and the compound eye are the most advanced types and provide a basis for discussing how the vertebrate eye could have developed, during the course of which other types of eye are described.

Figure 2.4   A fly's-eye image of Charles Darwin.

2.3. The Evolution of the Vertebrate Eye

Early life formed when conditions on Earth were very hostile to its existence. The atmosphere was originally devoid of oxygen but various processes were beginning to introduce oxygen into the atmosphere. One of the first life forms on Earth was *cyanobacteria*, a widespread green-blue bacterium that forms clumps in many locations, including bodies of water and damp rocks. There are suspected fossil records of this bacterium dating to 3.8 billion years ago. Like plants, they can perform photosynthesis and it is thought that they were responsible for first introducing oxygen into the atmosphere. With no oxygen available there was no ozone layer around Earth to shield the surface from harmful ultraviolet radiation and the cyanobacteria would have flourished only in sheltered environments such as in water or within damp rocky crevasses.

What was true for cyanobacteria would also have been true for later, slightly more complex, life forms while the ozone layer was sparse and formed an inadequate shield against harmful radiation. Any organism that could detect ultraviolet light and move into a shady environment that the light could not reach would clearly have had a greater chance of survival — the very condition required for Darwinian evolutionary theory to operate. Even bacteria have a device for locomotion — the whip-like *flagellae*, one or more of which are attached to its outer surface. At a slightly more advanced level the closing together of neighbouring surfaces can produce an outflow of water that would propel the organism in the opposite direction. However, whatever the means of propulsion, the first essential requirement is to detect where the light is coming from so that the required direction of motion is known.

The most primitive light detection system would have been photosensitive pigments on the surface of the organism (see Fig. 2.5(a)). This would have given only a very crude indication of the direction from which the light was coming, but better than nothing. An example of such a light-detection system in a primitive creature is found in *euglenids*, small unicellular organisms that form a green scum in pools of water. They are covered by a flexible membrane and contain chloroplasts, the chlorophyll within which performs photosynthesis to produce energy and also carbohydrates and other material necessary for its sustenance. For such an organism, light is an essential requirement for survival and it possesses a flagellum. At the base of the flagellum is a patch of red photosensitive pigment called an *eyespot* that has the function of controlling the flagellum so that it propels the organism *towards* light.

The next stage in the development of the eye is the existence of the light-sensitive pigments within a hollow recess, which would restrict the angle from which the light affecting the pigment could have come, hence increasing directional discrimination (shown in Fig. 2.5(b)). The *flatworm* has this kind of light-detecting system with two cups containing pigments forming primitive eyes on its head. However, this is basically the same system as that of euglenids, although with a better performance.

If the recess became deeper and the entrance to it more restricted then, with a single patch of pigment at the end of the recess, the directional discrimination would become extremely good. However, a new development is now possible: by having several patches of pigment the responses of different patches can be compared to give a sense of the distribution of light coming from various directions. If these patches become numerous enough, like individual photoreceptors, and they send separate signals to the brain then we have the

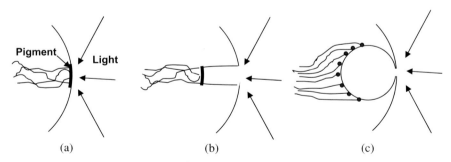

Figure 2.5   Three stages in the development of a primitive eye. (a) A pigment patch with poor directional discrimination. (b) Pigment within a hollow giving better directional discrimination. (c) A small aperture giving a pinhole-camera effect with several independent pigment patches.

Figure 2.6   Nautilus showing the 'pinhole-camera' eye (Berlin Zoo).

makings of image formation with a pinhole camera (Fig. 5.1). The cephalopod *Nautilus* (Fig. 2.6) has this type of eye, with a narrow aperture acting as a pinhole and with the interior of the cup open to the outside environment.

The simple types of eye described thus far have been fit for purpose for the limited requirements of the creatures that possessed them but would not suffice for the more complex lifestyles of advanced vertebrates. A modification in which the pinhole-aperture eye developed a surrounding membrane would protect the eye from outside contamination, and hence degraded performance, and the filling of the enclosed space with a clear fluid would then strengthen the structure of the eye. This would then have given a slightly improved eye, as represented in Fig. 2.7(a). The membrane would have to be clear in front of the pinhole aperture for the eye to operate and if this thickened in a non-uniform way, as illustrated in Fig. 2.7(b), then one has something like a lens being formed that would lead to both a brighter and a clearer image. With the addition of a lens, with a focal-length adjustment system and individual pigment patches converting into a large number of photoreceptors, one now has a structure similar to the modern vertebrate eye, as shown in Fig. 1.2.

The above somewhat simplified story of the evolution of the eye does not deal with complex issues such as the way that patches of pigment eventually became individual photoreceptors and how the

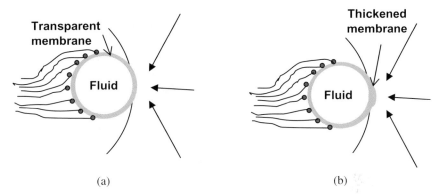

Figure 2.7 The beginning of the vertebrate eye. (a) A transparent membrane enclosing a fluid (b) A thickening of the membrane to produce a lens.

neural system developed in such a way as to interpret the increasingly complex signals it receives. However, what has been illustrated is that there is a pathway of ever-increasing complexity that goes from something very simple to something quite complex with each stage providing a light-detection or visual system of practical use to the organism that possesses it. In the animate world there are significant variations in eye structure and performance. The arrangement of the layers of retinal cells in the octopus is better than that in man and the visual acuity of birds of prey, which spot their quarry from great heights, is far greater than that of land-based mammals. It seems that, for any particular organism, facilities develop up to the point where it gains advantage from the development. If humans could see less well then they would be disadvantaged but a mutant man with the visual acuity of a hawk would be no better placed than his fellow men in terms of survival.

There may be difficult aspects of anatomy for Darwin's theory of evolution to explain, but the eye does not seem to be one of them.

Chapter 3

Waves and Image Formation

3.1. What is Light?

At the end of the seventeenth century there were two opposing views about the nature of light. The first, due to Isaac Newton (1643–1727; Fig. 3.1), was that light consisted of a stream of corpuscles, like bullets being fired from a machine gun. The second, due to the Dutchman Christiaan Huygens (1629–1695; Fig. 3.2), was that light was a wave motion, similar to the waves seen on the surface of a disturbed pool of water. At the beginning of the nineteenth century the wave theory became accepted due to the work of the English scientist Thomas Young (1773–1829) who showed interference effects with light, something that was also shown by water waves. Later the French scientist Augustin-Jean Fresnel (1788–1827) extended Young's work and established a sound theoretical foundation for light as a wave motion. Finally, the Scottish physicist James Clerk Maxwell (1831–1879) showed that light was an electromagnetic radiation — coordinated oscillating electric and magnetic fields moving through space with the speed of light.

In the study of optical image formation the wave theory explains all observed phenomena, and that is the approach we shall be using here. However, for completeness it should be noted that Newton's view of light was also correct. In 1887, the German physicist Heinrich Hertz (1857–1894) found that electrons could be expelled from a metal surface by a beam of light, a phenomenon known as the *photoelectric effect.* In 1902, the Hungarian–German physicist Philipp von Lenard (1862–1947; Nobel Prize for Physics in 1905 for his work on

Figure 3.1   Isaac Newton.

cathode rays) experimentally found the precise laws governing the photoelectric effect. Finally, in 1905, Albert Einstein (1879–1955) published a paper in which all Lenard's results were explained in terms of light consisting of corpuscles, known as *photons*. It was for this work that Einstein received the Nobel Prize for Physics in 1921.

Modern views of light are encapsulated in the concept of *wave-particle duality* in which light can behave either as a wave or as a particle depending on the observation being made. The concept of wave-particle duality also applies to other entities that we normally think of as particles, such as electrons, established as particles because they were found to be associated with discrete properties of mass and electric charge. In the following chapter the electron microscope is described, the working of which depends on the fact that electrons can have a wave-like behaviour and, like light, form an image.

Figure 3.2   Christiaan Huygens.

3.2. Huygens' Wavelets

Huygens proposed a model for the way that light moves away from a particular point on an object. At any instant the light has progressed to a spherical *wave front*, with the centre of the sphere being the emitting point. Now every point on the wave front becomes a source of wavelets so that after a time dt there is a new wave front, formed by the envelope of all the wavelets, at a distance $c\ dt$, where c is the speed of light. This is illustrated in Fig. 3.3.

It will be seen that the wavelets only create a new wave front in a forward direction. When Huygens described his model he gave no explanation of why a new wave front should not also be formed in a backward direction and that remained a difficulty of the model for some time. However, accepting that new wave fronts are only formed in a forward direction, this model gives a ready explanation of the

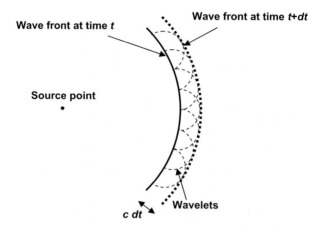

Figure 3.3   The wave fronts at times t and $t + dt$.

way that light behaves when being reflected and refracted and also the action of a lens in forming an image.

3.3. Reflection and Refraction

The principle that operates when considering reflection is that the wavelets cannot penetrate the reflecting surface and that part of a wavelet that would have projected beyond the surface appears instead in a reflected form. This is illustrated in Fig. 3.4 for the wavelet from the source that meets the reflecting surface. With this model of the way that wavelets behave at a reflecting surface we can now understand how a planar wave front is reflected. This is shown in Fig. 3.5 and indicates how the reflected wave front leaves the surface at a particular time, being tangential to all the wavelets.

The process of refraction occurs when light crosses a boundary on the other side of which the speed of light is different from that in the medium whence it came. This occurs when light moves from air to glass; the speed of light in common varieties of glass is about two-thirds of that in air. Figure 3.6 shows the refraction of a plane wave front of light moving from air into glass across a plane surface.

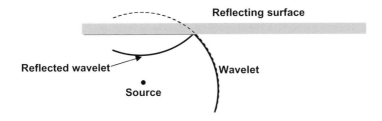

Figure 3.4   A wavelet reflected at a surface.

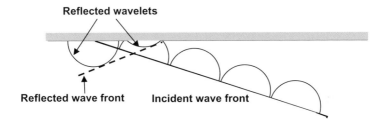

Figure 3.5   The formation of a reflection using Huygens' wavelets.

The wavelet at A moves into the glass and is represented by a hemisphere of radius smaller than that at B where the speed of light is greater. The wavelet from B is shown penetrating the interface and so the hemispherical surface will be distorted in the glass where the speed of light is lower.

Refraction is also the phenomenon that explains the behaviour of a lens in bringing rays emanating from a point to focus at a point. Figure 3.7 shows successive spherical wave fronts moving from the source point through the lens and then converging on the final image point, without showing wavelets that would make the diagram complicated and difficult to understand.

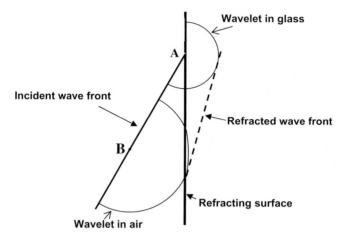

Figure 3.6   The Huygens' wavelet construction for refraction across a plane surface.

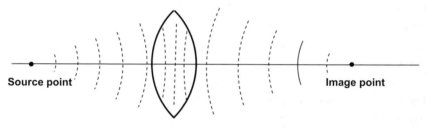

Figure 3.7   The wave fronts (dashed) from source point to image point through a lens.

While the Huygens wavelet model works well in explaining various effects in geometrical optics — reflection and refraction — it is not so successful in describing diffraction, an important aspect of physical optics. Fresnel modified the Huygens model in such a way that it explained both the phenomenon of diffraction and the reason why no backwards wave front was formed.

3.4. Stereoscopy

When we look at a photograph we are seeing an image of a scene in two dimensions. There are various aspects of the photograph that may contribute to a sense of depth, which means introducing a third dimension into our appreciation of what we see. Thus we know the relative sizes of various kinds of object, so a small house and a large man in the scene indicate that the man is much closer than the house. We expect the opposite edges of a road or path to be parallel so that when they appear to converge in our picture we know that where the sides are closer together the road is further away. Again a picture will be more focused in some parts than in others; this mimics the way that we view a natural scene because if our eyes are focused on near objects then objects further away are not in focus.

All the above clues to a third dimension in a photograph would be available to a one-eyed man. However, the majority of people have two eyes and this gives us stereoscopic vision in which, because each eye sees a slightly different aspect of an object, we obtain an impression of depth. This is the basis of taking a stereoscopic pair of images. In this technique two photographs of a scene are taken simultaneously with a special camera in which the two camera lenses are separated horizontally at about the distance of human eyes. There are two principal ways of observing a three-dimensional image from a stereoscopic pair of photographs. The first of these involves printing the images overlapping each other, one in red and the other in green. Such an image is shown in Fig. 3.8. To see the picture in depth, special viewing spectacles are required, the right one having a green lens and the left a red lens. In this way the right eye sees only the red image, seen as black on a green background, and the left eye the green image, seen as black on a red background.

In the second method the photographs are placed side by side as in Fig. 3.9. Stereoscopic viewing devices are available that ensure that each eye sees only one picture, that taken with the camera lens on the same side as the eye. However, many people can manage without the stereoscopic viewer. If the pair of images are placed 10 cm or so from the nose, a distance at which the eyes cannot focus on them, then the two blurred images are seen overlapped. If now they are slowly moved away, keeping the images overlapped, then eventually the eyes will bring the images into focus and the stereoscopic view will be achieved.

Figure 3.8   A red-green stereoscopic picture.

Figure 3.9   A stereoscopic pair of images.

The cinema industry has used the process of stereoscopy to produce three-dimensional films in full colour. In this they exploit the phenomenon of the polarization of light. Electromagnetic radiation is a *transverse wave motion*, meaning that the alternating electrical and magnetic fields of which they consist oscillate in a direction at right angles to the direction in which the wave is moving. Normal light, say that produced by a filament lamp, consists of short bursts of waves, all with their electric vibrations at randomly different angles to each other. This is unpolarized light. It is possible to produce glass or transparent plastics that contain materials that will transmit only the light polarized along one direction. If a slide, or a frame of a cine film, is projected with light that passed through such a *polarizer* then it would appear to be quite normal, although with a slightly reduced intensity since only the components of the vibrations of each light pulse parallel to the allowed direction would have been allowed through. If the projected image is viewed through a piece of polarizing glass with an acceptance direction parallel to that of the projection then it will be seen with undiminished intensity. However, if the polarizing glass is then rotated through a right angle then none of the light from the projection can pass through and the screen would appear to be completely dark. By projecting two images simultaneously, a right-eye image with horizontal polarization and a left-eye view with vertical polarization, and with appropriately polarized lenses of a pair of spectacles, the conditions for three-dimensional viewing are obtained.

3.5. Holography

The normal way of considering how we see an image is to think of it in terms of geometrical optics where rays of light emanating from a point on an object are brought together by a lens or mirror system to give an image point. Another, and somewhat more complicated, way is to think in terms of wave fronts; in the case of viewing an object with the eye, the lens in Fig. 3.7 is that of the eye. In Fig. 3.10 we show a wave front for the light passing from an object to the eye. If the eye moves then the same wave front can show the object in a different perspective.

Each point on the wave front is receiving light from each point of the object and at any instant, with white light, at each of these points light of a particular wavelength is characterized by a particular amplitude and phase. The amplitude describes the maximum extent

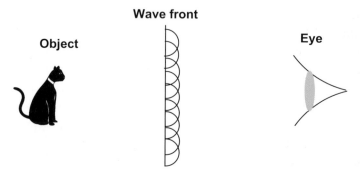

Figure 3.10   An intermediate wave front in the process of forming an image.

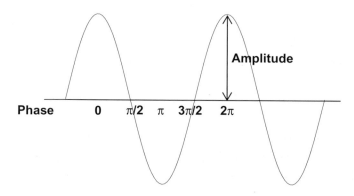

Figure 3.11   The amplitude and phase of a wave.

of the vibration that constitutes the electromagnetic wave and the phase the distance from the crest of a wave, expressed as an angle where from crest to crest is represented by 2π (Fig. 3.11). The intensity of the radiation is proportional to the square of the amplitude. If some process existed whereby one could precisely re-create the wave front in the absence of the object, then an image of the object would be seen, albeit that the object was not actually present. This would mean that at every point the amplitude and phase for every wavelength would have to be precisely that produced by the object. With normal light there is no way of achieving this objective. However, in 1958, the American scientists Arthur Schawlow (1921–1999) and Charles Townes (b. 1915; Nobel Prize for Physics, 1964) invented the laser (Light Amplification by the Stimulated Emission of Radiation), a new kind of light source. A substance, which can be a solid, gas or vapour, is excited electrically so that the electrons of a specific type of atom are raised to a higher energy level. The substance is contained within a pair of parallel plane mirrors that can reflect light to and fro. Initially electrons spontaneously fall to a lower energy level, emitting photons of specific wavelengths. When one of these strikes an excited atom it stimulates the emission of a photon of precisely the same energy and in phase with the colliding photon. Eventually the only mode of photon production is of stimulated photons moving along the axis of the device precisely perpendicular to the mirrors. If one of the mirrors is partially transmitting then from that mirror a very narrow parallel beam of radiation is emitted with very high intensity. The emitted beam has the characteristic that it consists of very long continuous wave trains with all photons precisely in phase — a beam of *coherent radiation.*

In 1964 the Hungarian scientist Dennis Gabor (1900–1979; Nobel Prize for Physics, 1971), who worked in the UK, used the laser to achieve the goal of preserving the information of a wave front from an object and later recreating the wave front to produce a three-dimensional image of the object without the object being present. This technique, known as *holography*, originally required very precise experimental conditions to achieve it but is now so robust that postage stamps have been issued displaying *holograms*. One possible arrangement for producing a hologram is shown in Fig. 3.12.

A plane wave strikes a partially silvered mirror so that some passes straight through and illuminates the object while the remainder is reflected towards another mirror and from thence to a photographic plate. This part of the illumination of the plate is called the *reference beam.* Some of the light scattered by the object also falls on the photographic plate. Since the light is coherent the phase difference in the light coming via the two routes is always the same, depending on the difference in path lengths, and is independent of time. If they are in phase then their resultant will be large and the photographic plate blackened but if they are out of phase the resultant will be small and the plate will not be greatly affected.

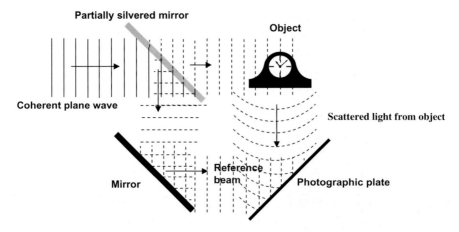

Figure 3.12   An arrangement for producing a hologram.

When the photographic plate is developed it shows different intensities at different places (Fig. 3.13). If the light were not coherent the plate would have a uniform exposure and contain no information about the object. Rather surprisingly, this two-dimensional blotchy plate contains sufficient information to give an image of the object with the possibility of seeing it in different perspectives, by shifting the viewpoint over a limited range.

The arrangement for viewing the object is shown in Fig. 3.14.

The photographic plate is illuminated by a laser similar to that which produced the hologram. Each point of the hologram scatters light with an intensity that depends on how blackened it is at that point; the greater the blackening the less is transmitted and the more scattered. The eye, looking in the direction shown, receives wave fronts formed by this scattered light apparently coming from a virtual image of the original object.

The actual physics of the hologram is best explained in mathematical terms, most efficiently using a wave described in terms of complex numbers. An analysis follows for those readers with the necessary mathematical background and who wish to understand the principle of holography in more detail.

Figure 3.13   A developed hologram.

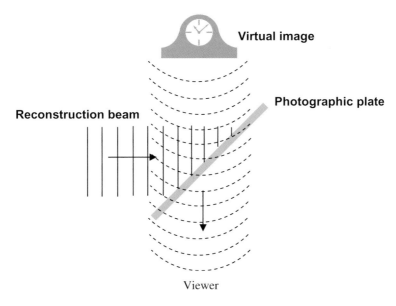

Figure 3.14   Arrangement for viewing a hologram.

We begin by expressing a wave as

$$y = A \exp\left\{2\pi i \left(\frac{x}{\lambda} - \nu t\right)\right\} = A \exp\{f(x, t)\} \qquad (3.1)$$

in which y is the displacement at position x and time t, A is the amplitude of the wave and λ and ν are the wavelength and frequency of the light. The amplitudes of the light arriving from the reference beam and the object at a point on the photographic plate can be described, respectively, as

$$y_R = A_R \exp\{f(x, t)\} \qquad (3.2)$$

$$y_O = A_O \exp\{f(x, t) + \phi\} \qquad (3.3)$$

The amplitude A_R is constant and its square represents the uniform intensity of light on the plate if the reference beam were present alone. The displacements of both of the two arriving waves are changing with time but always maintain the same difference of phase, ϕ depending on the difference of path length to the plate. The intensity at the point on the photographic plate is

$$I = |y_O + y_R|^2 = (y_O + y_R)(y_O^* + y_R^*) = |y_O|^2 + |y_R|^2 + y_R y_O^* + y_O y_R^* \qquad (3.4)$$

where y^* is the complex conjugate of y.

Substituting from (Eq. 3.3) into (Eq. 3.4):

$$\begin{aligned}
I &= A_O^2 + A_R^2 + A_O A_R \exp\{-f(x, t) - \phi\} \exp\{f(x, t)\} \\
&\quad + A_O A_R \exp\{-f(x, t)\} \exp\{f(x, t) + \phi\} \\
&= A_O^2 + A_R^2 + A_O A_R \{\exp(-\phi) + \exp(\phi)\} \\
&= A_O^2 + A_R^2 + 2 A_O A_R \cos\phi \qquad (3.5)
\end{aligned}$$

If the blackening of the plate is proportional to I then the fraction of the reconstruction beam transmitted through the plate is $1 - kI$, where k is a constant and the fraction kI is scattered by the blackened emulsion of the plate. A fraction of this will enter the viewer's eye

(Fig. 3.14) so that the displacement of the radiation received from a particular point of the photographic plate, which varies with time, is

$$Y = C \exp\{f(x,t)\}(A_O^2 + A_R^2 + 2A_O A_R \cos \phi)$$

that can be expressed as

$$Y = C(A_O^2 + A_R^2)\exp\{f(x,t)\} + CA_O A_R \exp\{f(x,t) + \phi\}$$
$$+ CA_O A_R \exp\{f(x,t) - \phi\} \tag{3.6}$$

The first term on the right-hand side has no information content and is just a general background illumination. Because A_R is constant, the second term is like y_O in Eq. 3.3 — the light coming from the original object. Consequently, this term forms a three-dimensional virtual image of the original object shown in Fig. 3.14; changing the viewing direction changes the perspective of the original object that is viewed. The third term in (Eq. 3.6), with phase reversed, forms what is called the *conjugate image*, a real image that is not in the same place as the virtual image. The viewing of the virtual image is degraded by the presence of the background and conjugate-image radiation but there are ways of arranging holograms so that the various components in (Eq. 3.6) do not overlap so that a clear image is seen.

Chapter 4

Seeing Small Objects

4.1. Resolution of the Visual System

When we look at an ant scurrying about in the garden we see that it is a small creature a few millimetres in length, that it is divided into a head and a body consisting of two connected bulbous parts and that it has six legs, but we cannot see more than that. If we place the ant on a piece of paper and bring it in to the minimum distance at which we can focus our eyes — the *near point*, usually about 25 cm — then we may see a little more detail, but not much more. What defines the limit of what we can see? Well, clearly the packing of the photoreceptors in the retina provides one constraint. The diameter of the eye is about 2.5 cm, just one-tenth of the near-point distance, so an object 5 mm long seen at the near point would form an image on the retina of size 0.5 mm. The diameter of a cone is 0.006 mm so, in the fovea where they are most tightly packed, the image would span 83 cones — a clear limitation of resolution.

The structure of the retina is not the only determinant of resolution; the wave nature of the radiation we are using and the size of the imaging equipment, e.g. the eye, also play a role. To see how this matters we first look at the diffraction (scattering) of a parallel beam of light of wavelength λ falling on an aperture in the form of a slit with width d (Fig. 4.1). Each point of the aperture diffracts radiation in all directions (remember Huygens' wavelets) and in the forward direction all the radiation being scattered from different parts of the aperture are in phase and add up to give a large intensity. Now we consider the radiation scattered at an angle θ from a point A at the

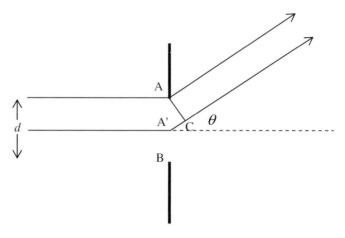

Figure 4.1   The diffraction of light from a slit of width d.

edge of the aperture and from a point A′ in the middle of the aperture. If the distance A′C is $\lambda/2$, half a wavelength, then wave crests and troughs from A are cancelled out by troughs and crests from A′ and they give no net contribution in the direction indicated by θ. To every point within the aperture between A and A′ there corresponds another point between A′ and B with the same cancelling relationship so there will be zero amplitude, and hence zero intensity, in the indicated direction. From the figure it will be seen that the angle between CA and AA′ is θ and hence

$$\sin\theta = \frac{A'C}{AA'} = \frac{\lambda/2}{d/2} = \frac{\lambda}{d}. \tag{4.1}$$

For the direction θ in Fig. 4.1, the path difference between rays leaving A and B is λ so that radiation from the region between A and A′ balances out that from the region between A′ and B. If the path difference in radiation from A and A′ is an odd number of half-wavelengths, i.e. $(2n+1)\lambda/2$, where n is an integer, then again there will be pairs of zones of the slit cancelling each other. Hence the condition for a minimum (zero) intensity is that radiation coming from A and B should have a path difference of $(2n+1)\lambda$, where n is

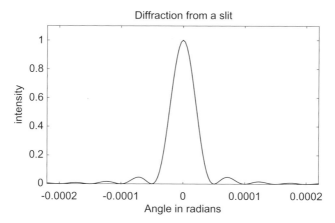

Figure 4.2   The intensity distribution with angle of the diffraction of light of wavelength 500 nm from a slit of width 1 cm.

an integer, so that there are minima for

$$\sin\theta_{\text{min}} = \frac{(2n+1)\lambda}{d}. \qquad (4.2\text{a})$$

Between the minima there will be maxima close to, although not precisely at, where the path difference between A and A′ is a whole number of wavelengths, when the angle θ is given by

$$\sin\theta_{\text{max}} = \frac{2n\lambda}{d}. \qquad (4.2\text{b})$$

The variation of the intensity of the diffracted light with angle is shown in Fig. 4.2. The light that consisted of parallel rays when it entered the slit, say the light from a star that is a point source a great distance away, is now spread out over a small angle. In the figure the slit is 1 cm wide and the wavelength of the light is 500 nm, which is in the green part of the spectrum. The first minimum is separated from the peak by 5×10^{-5} radians,[1] or just over ten seconds of arc.

Now we consider the overall intensity pattern where there are two equal sources, say two stars, subtending a small angle at the very distant viewing position. This is shown in Fig. 4.3(a) for an angular

---

[1] A radian is a unit of angle equal to 57.2958°. It is the angle subtended at the centre of a circle by an arc on its circumference equal in length to the radius.

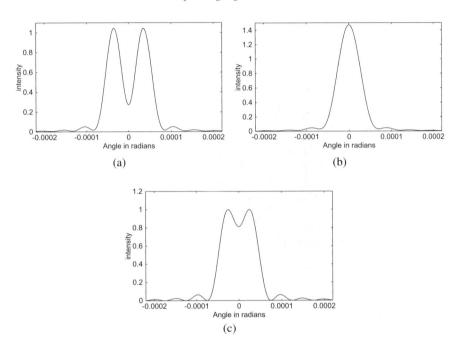

Figure 4.3   The overall intensity for the diffraction with a slit of width 1 cm using radiation of wavelength 500 nm for two objects with angular separations in radians (a) 0.00007 (b) 0.00003 and (c) 0.00005.

separation of 0.00007 radians. Two distinct peaks are seen and the images are resolved. However, for an angular separation of 0.00003 radians the intensity pattern, as seen in Fig. 4.3(b), shows only a single peak and hence the images are not resolved. Figure 4.3(c) shows the result for an angular separation of 0.00005 radians, which corresponds to the peak of each image coinciding with the first minimum of the other one and the images can just be resolved. This is the so-called *Rayleigh criterion* for resolution, named after Lord Rayleigh (1842–1919), an English physicist who received the Nobel Prize for Physics in 1904 for the discovery of the element argon. The angular separation that can just be resolved is λ/d and clearly is less if d is greater.

Resolution is important for astronomical telescopes for which the quantity playing the role of the slit width in determining the limit of resolution is the diameter of the objective lens or mirror (§§8.2

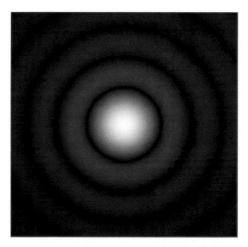

Figure 4.4   Diffraction with a point source and a circular aperture showing the Airy disk.

and 8.3). For a circular aperture, such as a lens, the diffraction pattern of a point source, such as a distant star, consists of a series of light and dark rings (Fig. 4.4). The central bright region is known as the *Airy disk*, named after the English astronomer Sir George Biddell Airy (1801–1892). The condition for the limit of resolution of the images of stars for a telescope is

$$\sin\theta = \frac{1.220\lambda}{D}, \qquad (4.3)$$

in which D is the diameter of the objective lens or mirror of the telescope. For a large astronomical telescope with a mirror of diameter 8 m and taking λ as 500 nm the smallest angular separation of two objects that can be resolved is 0.016''.[2]

Returning to the eye, the limitation of angular resolution attainable due to the structure of the retina is the distance between photoreceptors, 0.006 mm, divided by the distance of the retina from the eye lens, 25 mm. In terms of angle this amounts to about 50''. The pupil of the eye, the aperture through which the light enters, is

---

[2]The symbol '' represents 'seconds of arc'.

typically 4 mm in diameter and applying (Eq. 4.3), for light of wavelength 700 nm which is the greatest wavelength it can detect, gives an angular resolution of 44″. This is very similar to the limitation due to retinal structure — evolution seems to have provided a balance.

Resolution is always an important consideration in imaging systems. In the remainder of this chapter we shall be considering the imaging of small objects, both with visible, and near-visible, light and by using electrons, by the process of microscopy.

4.2. A Simple Microscope — the Magnifying Glass

The idea of using a lens as a magnifying device is probably quite old. A raindrop can act as a very crude lens and a blob of dried resin from a tree can do likewise and be acquired as a possession to be used anywhere as required. The oldest lens artefact known is from Assyria, the region of modern Iraq, and is the Nimrud lens made of rock crystal (pure quartz), dating back about 3,000 years. It was probably used by Assyrian craftsmen in making extremely fine engravings. Spectacles, as aids to vision, seem to have been invented in Italy during the thirteenth century and can be seen in some oil paintings from the early fourteenth century. We are all familiar with the Sherlock Holmes character, either in films or in book illustrations, searching for clues with a magnifying glass. Indeed many of us, particularly older people, keep a magnifying glass in the home to examine the fine print that sometimes is used on product labels or on official documents. The warning 'Read the small print' is usually difficult without such a visual aid. The common magnifying glass is sometimes referred to as a *simple microscope*, which enables us to see an enlarged version of the object of interest.

If our eyes could focus on an ant at a distance closer than the usual near point of 25 cm then the object would subtend a greater angle at the eye lens and the image would be larger on the retina and cover many more photoreceptors, thus increasing the fineness of detail we could discern. One arrangement by which this may be done is shown in Fig. 4.5. The object is placed at a distance from

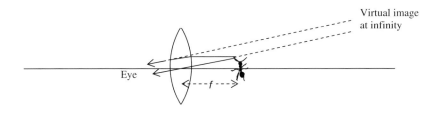

Figure 4.5   The action of a simple microscope.

the lens equal to its focal length, f. The parallel rays that come from
the lens from each point of the object appear to have come from
a virtual image at an infinite distance, one that subtends the same
angle at the eye as the object at distance f but one that can be seen
comfortably by the normal eye. If the eye were close to the lens and
the focal length of the lens, f, is less than the near-point distance
then, the effective magnification will be (near-point distance)$/f$. For
a lens of focal length 5 cm and a normal eye this will give a five-
fold magnification. With a lens of very short focal length and of
high quality a magnification up to about 20 is possible for a simple
microscope.

It is important not to confuse magnification with resolving power;
a large image that is blurred does not give a higher resolving power.
A critical factor in the resolving power of a simple microscope is the
diameter of the lens in relation to its focal length. If the lens is very
small then it will collect little of the light scattered by the object
being viewed; the more light that is collected the more information
there will be about the fine detail of the object. The question of
how resolving power is related to the diameter of the lens being used
was developed by Ernst Abbe (1840–1905), a German physicist who
developed many optical instruments and was a co-owner of Carl Zeiss
AG, a famous manufacturer of optical instruments of all kinds. We
can get the essence of his theory of resolving power by considering
the light entering a lens that has been scattered by a pair of parallel
very narrow slits with separation d that is much greater than the
width of the slits (Fig. 4.6).

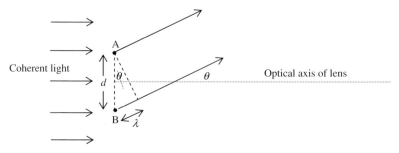

Figure 4.6   The formation of the first diffraction maximum from a pair of slits A and B.

The light falling on the slits is coherent, meaning that for every wave train of light falling on slit A there will be a similar wave train, with the same phase, falling on B. The light scattered in the forward direction by each slit will clearly be in phase so will reinforce to give a strong beam in that direction. The figure shows light scattered at an angle θ such that the light forming the wave front in that direction is also strong because the path length difference from A and B is just one wavelength. This is known as the *first-order diffracted beam.* There will be another first-order diffracted beam on the other side of the optical axis. From the figure we see that

$$\sin \theta = \frac{\lambda}{d}. \tag{4.4}$$

For the lens to be able to form a resolved image of the slits it is necessary that it should receive both the zero-order beam (the one going straight forward) and a first-order beam. When the distance between the slits is very tiny compared to the size of the lens imaging them we may say that for resolution of the two slits, distance d apart, the semi-angle that the lens must subtend at the object (the pair of slits in this case) is θ. Expressing this in another way, if the lens subtends an angle 2θ at the object then the smallest distance it can resolve is

$$d = \frac{\lambda}{\sin \theta}. \tag{4.5}$$

In this context the quantity $\sin \theta$ is called the *Numerical Aperture* (NA) of the lens. Since $\sin \theta$ cannot be greater than unity, and from geometrical considerations must be less, the smallest distance that can be resolved is greater than λ, the wavelength of the light being used. For a magnifying glass used in the normal Sherlock Holmes manner, with NA about 0.25, the smallest distance that can theoretically be resolved[3] is of order $2\,\mu$m but the lens would have to be of extraordinarily high quality to achieve this resolving power.

4.3. The Compound Microscope

The origin of the idea of using lenses in combination to create what we might think of as an optical instrument is not known for certain but a German lens-maker Hans Lippershey (1570–1619), working in Holland, seems to have been the inventor of the *compound microscope*, a device with two or more lenses that magnifies small objects. A very famous example of an early microscope was that designed and constructed by the eminent English scientist Robert Hooke (1653–1703). This instrument (Fig. 4.7) was used to

Figure 4.7   Robert Hooke's microscope.

---

[3] $1\,\mu$m (micron) is 10^{-6} m.

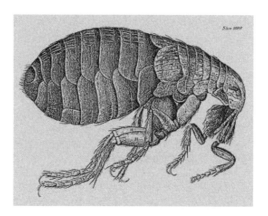

Figure 4.8   Hooke's drawing of a flea.

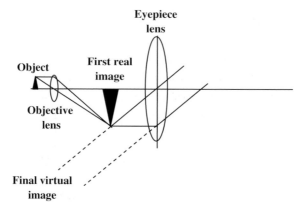

Figure 4.9   A basic compound microscope.

reveal tiny objects in detail never previously seen. Figure 4.8 shows Hooke's drawing of a flea, published in his groundbreaking book, *Micrographia*.

Figure 4.9 shows the action of a basic compound microscope. The objective lens has a very small focal length, usually a few millimetres, and there are very precise focusing devices to ensure that the object is correctly placed to produce the first highly enlarged real image. The eyepiece lens then acts like a simple microscope to view this real image, which is at or near a distance from the eyepiece equal to its focal length. The magnifying power of the whole instrument is the product of the magnification produced by the objective, usually in the range 5 to 100, and that produced by the eyepiece, usually in the range 10 to 15. Thus for the best instruments there is a magnifying power of 1,500 (100×15).

As previously mentioned, magnification is only a benefit to the extent that the fine detail of the object can be resolved and for a given wavelength of radiation this depends on the numerical aperture of the lens — in the case of a compound microscope that of the objective. There are two ways that the limitation of resolving power indicated in (Eq. 4.5) can be overcome. In Fig. 4.9 it has been assumed that the space between the object and the lens is occupied by air. However, if a liquid with high refractive index, say water or oil, is introduced then this can increase the effective NA of the lens and hence improve the resolution. To understand this we need to consider the way that light is refracted, i.e. bent, when it passes from one medium to another. Indeed, it is the process of refraction that enables a lens to focus light. In Fig. 4.10 a ray of light is shown passing from a medium of refractive index μ_1 into one of refractive index μ_2. The angle i is the *angle of incidence* and r the *angle of refraction*. These angles are connected by *Snell's law*

$$\frac{\sin i}{\sin r} = \frac{\mu_2}{\mu_1}. \tag{4.6}$$

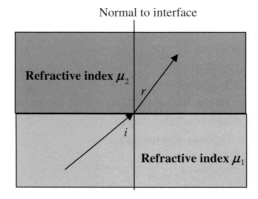

Figure 4.10   The passage of a light beam from one medium to another.

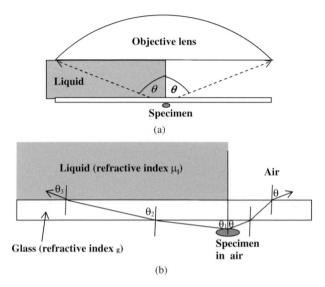

Figure 4.11   Comparison of extreme light paths from specimen to objective with and without intervening high refractive index liquid. (a) General view. (b) Close-up near specimen.

The refractive index of air is 1.00, that of water 1.33 and those of different glasses are variable but usually about 1.5. Commonly used oils have refractive indices in the range 1.5 to 1.6.

Figure 4.11 shows a comparison of the situation when the space between the glass cover slip, which protects the specimen, and the objective lens is air (right-hand side) and when it is occupied by a high refractive index liquid (left-hand side). In Fig. 4.11(a) the rays are shown between the glass cover slip and the edge of the objective lens but not the detail of the rays from the specimen to the top of the cover slip, which is shown in Fig. 4.11(b). To a good approximation $\theta_3 \equiv \theta$ since the lateral spread of Fig. 4.11(b) is small compared with the diameter of the objective lens. The effective NA, which is a measure of how much scattered light from the specimen reaches the objective, is $\sin \theta_1$. Using Snell's law

$$\frac{\sin \theta_1}{\sin \theta} \approx \frac{\sin \theta_1}{\sin \theta_3} = \frac{\sin \theta_2}{\sin \theta_3} \frac{\sin \theta_1}{\sin \theta_2} = \frac{\mu_l}{\mu_g} \frac{\mu_g}{1} = \mu_l. \tag{4.7}$$

Hence the NA for this arrangement is $\sin \theta_1 = \mu_l \sin \theta$.

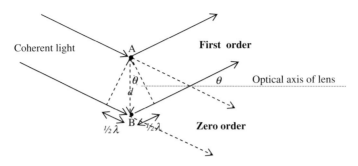

Figure 4.12   Imaging a pair of slits with oblique illumination.

The second way of increasing resolution depends on the way that the specimen is illuminated. In deriving (Eq. 4.5) it was assumed that the zero-order beam and *both* first order beams entered the lens. However, to attain resolution of the slits in Fig. 4.6 it is only necessary for the lens to intercept the zero-order beam and *one* of the first-order beams. Figure 4.12 shows an arrangement of illumination in which the zero-order beam passes through an edge of the lens and one of the first-order beams passes through the other end of the diameter of the lens. Now half the angle subtended at the slits by the lens is given by

$$\sin \theta = \frac{\frac{1}{2}\lambda}{d} = \frac{\lambda}{2d}, \tag{4.8}$$

which is one-half of the previous value. With these two improvements the resolution attainable with a microscope is given by

$$d = \frac{\lambda}{2\mu_l \sin \theta}. \tag{4.9}$$

The description given here of a compound microscope is at a very basic level. The optical systems of the actual instruments that are used today are, by comparison, very complex. The refractive index of materials is dependent on the wavelength of the light passing through them so a simple lens made of a single material would bring different wavelengths into focus at different points — a fault known as *chromatic aberration*. This is corrected by having *achromatic lenses*, combinations of lenses made of different materials that

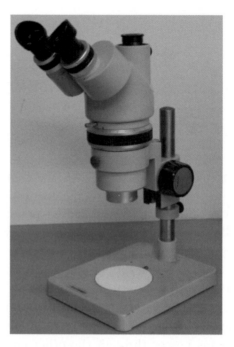

Figure 4.13   An optical stereoscopic microscope.

together act like a simple lens with the same focal length for all visible wavelengths. It will be seen in Fig. 4.9 that the final virtual image is inverted that, in some cases, is undesirable. For this reason microscopes include an *inverting lens* that takes no part in the magnification but simply inverts the first real image. Finally, there are stereoscopic microscopes with two eyepieces that look at slightly different perspectives with each eye and so give a stereoscopic view of the specimen (Fig. 4.13).

Another aspect of microscope design on which we have only lightly touched is the importance of proper illumination of the specimen. As was shown above, by directing the illumination appropriately it is possible to improve the resolving power. It is also important to have a well-illuminated specimen as this also assists in seeing fine detail.

A different way of improving the resolving power of a microscope, as will be clear from Eq. (4.9), is to use a smaller wavelength of radiation. Normally white light is used, in the visible wavelength range of 400 to 700 nm, with achromatic optical components to avoid chromatic aberration. It is possible to use ultraviolet light, in the range 200 to 400 nm for microscopy, giving an improvement by a factor of two or more in resolution over that attained with visible light. Most normal glasses are not very transparent for ultraviolet light far from the visible range. For that reason ultraviolet microscopes require the use of lenses made of quartz or fluorite, minerals that transmit ultraviolet light quite well down to about 200 nm. Another problem is that the images produced by an ultraviolet microscope cannot be seen directly. There are fluorescent materials that produce a visible image when the ultraviolet image is projected onto them. In addition, photographic materials are sensitive to ultraviolet light and so the images produced can be recorded and then viewed via photography.

4.4. Phase-Contrast Microscopy

Many of the objects studied with microscopes are transparent and absorb little of the light that passes through them. This makes them almost invisible and the problem is made greater if there is detailed structure in the object, the different parts of which transmit almost all the light falling on them. In such a situation it is very difficult to see the structure because the field of view is virtually uniform with only tiny variations of intensity. The Dutch physicist Frits Zernicke (1888–1966) solved this problem in 1930. His solution, known as *phase-contrast microscopy*, earned him the Nobel Prize for Physics in 1953. What Zernicke realised was that, although the radiation in the image plane defining the different parts of the object had a nearly uniform intensity, it varied markedly in phase from one image point to another, forming a *phase-object*. The technique devised by Zernicke was one that converted this difference of phase into a difference of intensity so as to introduce contrast in the image.

The phase differences that occur in different parts of the image are due to differences of thickness and refractive index of different parts of the object. If the speed of light in a vacuum is c then the speed in a material of refractive index μ is c/μ. The relationship between the speed of light, its wavelength, λ, and its frequency, ν, is given by

$$c = \lambda\nu, \tag{4.10}$$

and, since the frequency does not change, the wavelength within a medium of refractive index μ is λ/μ, where λ is the wavelength in a vacuum. The number of wavelengths fitting into a distance d is therefore d/λ in a vacuum and $\mu d/\lambda$ in the medium. This introduces a phase delay of the light passing through the medium compared with light passing through the same distance of a vacuum of

$$\Delta\phi = 2\pi\frac{d}{\lambda}(\mu - 1) \text{ radians.} \tag{4.11}$$

For a specimen of thickness $5\,\mu$m, light of wavelength $5.5 \times 10^{-7}\,$m and $\mu = 1.35$, $\Delta\phi = 2\pi \times 3.182 \equiv 2\pi \times 0.182 = 1.14$ radians.[4] Normally we are concerned with differences of phase in different parts of the specimen. If two neighbouring regions have the same thickness, $5\,\mu$m, but refractive indices 1.35 and 1.34 respectively then the light passing through the former would lag in phase behind that coming through the latter by

$$\Delta\phi = 2\pi\frac{5 \times 10^{-6}}{5.5 \times 10^{-7}}(1.35 - 1.34) = 0.57 \text{ radians,} \tag{4.12}$$

about $33°$.

Phase differences between 0 and π centred on $\pi/2$, are common from one part of a biological specimen to another. The arrangement for achieving the phase contrast is illustrated in Fig. 4.14. A compact light source, L illuminates an annular ring, R, which lets through a cone of light. This light passes through a condenser lens, C, that brings it to a focus to illuminate the specimen, S, located at the apex of the focused cone. The specimen contains components of different thickness and refractive index that scatter some of the light inwards and outwards away from the surface of the cone. The light scattered

---

[4]The effect of phase differences is not affected by whole numbers of 2π.

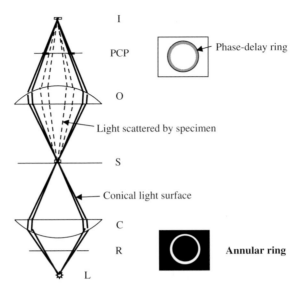

Figure 4.14   A schematic phase-contrast microscope. Only the light not scattered by the specimen passes through the deposited material on the phase-delay ring.

both within and outside the cone then passes through the objective, O. Positioned at P is the phase-contrast plate, PCP. This is the plane in which the combined optical components C and O produce an image of the annular ring, R. The phase-contrast plate is a uniform glass plate on which is deposited a thin layer of material in the form of an annular ring that precisely occupies the position of the image of R and imposes a phase delay, usually of $\pi/2$, on light passing through it. The scattered light from the specimen does not traverse the phase shifting ring and so is unaffected by the deposited material. This light forms a primary image, I, of the specimen that is overlaid by radiation with a phase shift $\pi/2$ that has passed though the phase-plate ring.

For the best result it is important that the light passing through the ring should not be of an intensity that overwhelms the image so, for this reason, the ring not only changes the phase of the light passing through it but also reduces its intensity by 70 to 90%.

There are several ways of explaining the working of a phase-contrast microscope. One way is first to consider the formation of an image by a lens, which we shall call the 'true image', taking the lens

to be formed by the sum of a large number of thin annular rings. Each
of these rings, taken alone, would form an imperfect blurred version
of the true image but the light forming each blurred image comes
from the same source and so is coherent from one to the other. The
lens sums these coherent blurred images by combining amplitudes,
taking into account phase differences, and together they form the
true image at a resolution determined by the NA of the lens.

The light coming from the specimen within the conical region,
which is scattered in a forward direction, passes through an annular
ring of the objective lens and so, if taken alone, would give a blurred
image, which, in amplitude and phase, will at each point be some
average of what is present in a small surrounding region of the true
image. Now let us consider how this, with amplitude reduced as by
the phase-delay ring but with *no* phase shift, would combine with
the good quality image from the remainder of the lens. We show its
contribution in Fig. 4.15(a) on two neighbouring points of the true
image with similar amplitudes but different phases. Because of the
averaging process the amplitude and phase of the blurred image will
be roughly the same at the two points with phase an average of the
true phases at those points and hence make the same angle with

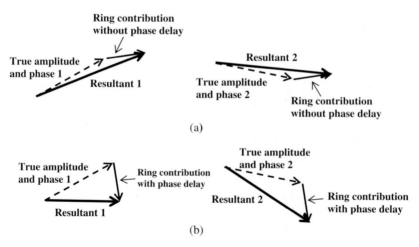

Figure 4.15 Adding the annular ring contribution to the image. (a) Without
phase shift and (b) with phase shift.

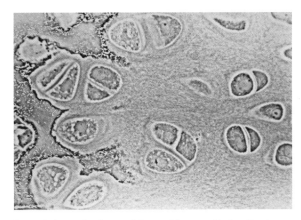

Figure 4.16   An epiphyseal plate with cartilage cells in the lower part showing calcification.

the true phases. Thus, we can see from the figure that when the blurred image combines with the contribution at those points the phase difference with the true phase is similar in magnitude at each of them. Hence, at those points the resultant amplitude is similar in each case. Next, in Fig. 4.15(b) we add the phase shift of the phase-delay ring, which has changed the phase of the blurred image by $\pi/2$. The resultant amplitudes are now very different and the intensities (squares of the amplitudes) of the two points of the image will also be very different.

Figure 4.16 is an example of a phase-contrast image. The way in which bones grow in children and adolescents is that there is a region towards the end of each bone where cartilage is formed that later calcifies and transforms into bone. This cartilage is produced in a plate-like form, *epiphyseal plates*, and the figure shows a plate in which some of the cartilage cells are calcifying.

There are ways of obtaining phase-contrast microscopy other than the one described here but this one illustrates the basic theory behind all of them.

4.5. Electron Microscopy

Although Albert Einstein (1879–1955) is best known for his seminal work on relativity theory he won his Nobel Prize for Physics in 1921 for a paper written in 1905 on the photoelectric effect, the phenomenon of the emission of electrons from metals when they are illuminated by light. Einstein's explanation of this phenomenon was that in this instance light, which is normally thought of as a wave-like entity, was behaving like a stream of particles, each particle with an energy and momentum characteristic of its wavelength. In 1924, in a doctorate thesis, a young French student, Louis de Broglie (1892–1987; Nobel Prize for Physics, 1929) put forward the hypothesis that the converse phenomenon might also happen — that entities usually thought of as particles might sometimes evince wave-like behaviour. The formula he gave that connects the particle and wave properties is

$$\lambda = \frac{h}{p} \tag{4.13}$$

where λ is the effective wavelength of the particle, p its momentum and h is Planck's constant.[5] In 1927 two American physicists, Clinton Davisson (1881–1958; Nobel Prize for Physics, 1937) and Lester Germer (1896–1971) carried out an experiment in which they diffracted a beam of electrons from a crystal, in a similar way that x-rays are diffracted from crystals (Chapter 15), so confirming the de Broglie hypothesis.

The resolution possible with any microscope is of the same order as the wavelength of the radiation, or other type of wave motion, being used for the imaging (Eq. 4.5). In the case of an ultraviolet microscope this will be of order 200 nm. However, for an electron of energy 1,000 eV, i.e. that of an electron accelerated through a potential difference of 1,000 V, the de Broglie wavelength is 4×10^{-11} m, some 5,000 times less than that of short-wavelength ultraviolet light. In 1931 two German scientists, the physicist Ernst Ruska (1906–1988; Nobel Prize for Physics, 1986) and electrical engineer Max Knoll (1897–1969) designed the first electron microscope to exploit the wave property of electrons. By 1933 their instrument exceeded the resolving power of optical microscopes and since then development

---

[5]Planck's constant, $h = 6.626 \times 10^{-34}$ Js, occurs in all aspects of quantum theory.

has continued so that now magnifications of up to 2 million have been achieved with resolution less than 1 nm.

A microscope must have a number of essential components. In the case of an electron microscope these are:

- A source of electrons, a so-called *electron gun*, with a tight energy distribution, corresponding to having monochromatic radiation for an optical microscope.
- Lenses to focus the electrons and form an image. In the case of the electron microscope, lenses can be designed using either electric or magnetic fields.
- Some means of viewing the image.

The simplest form of electron gun is a heated metal filament, which can be either made of tungsten, which is very durable, or of some other material that produces electrons more readily but may degrade much faster. In another form of gun a filament is subjected to a very intense electric field, giving what is called a *field emission gun* (FEG), which can have either a warmed filament, to assist the escape of electrons, or a cold filament. The latter of these, the cold FEG, requires the most demanding technology but it also gives the smallest energy spread in the emitted electrons, about 0.25 eV. At the other end of the scale with a heated filament the energy spread can be as much as 2.5 eV. For the most precise work in terms of resolution, a cold-filament electron gun would be the obvious choice, albeit that they have a limited lifetime and are expensive to maintain.

Since electrons are negatively charged particles, lenses based on either electric fields or magnetic fields can be used to focus them and produce an image. The action of both electric and magnetic fields is to produce a force on an electron, although in the case of a magnetic field the electron must be moving for the force to be generated. An electron, of charge e (coulombs), in an electric field E (volts per metre) will experience a force eE (newtons) in the direction of the field. For a magnetic field B (tesla) and an electron moving with speed v (metres per second) the force, both in magnitude and direction, depends on the directions of B and v. The relationship is shown in Fig. 4.17. With the angle θ between B and v the force on

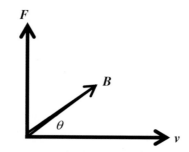

Figure 4.17   The force on a moving charge due to a magnetic field.

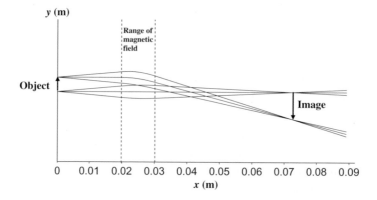

Figure 4.18   The action of a magnetic electron lens.

the electron is perpendicular to the plane defined by B and v with magnitude

$$F = Bev \sin \theta. \tag{4.14}$$

In the figure, B is pointing away from the reader.

We can illustrate the principle of operation of a magnetic lens by a computational example. In Fig. 4.18 we show the effect of a magnetic lens on electrons coming from an object in the form of an arrow at $x = 0$, extending from the origin to $y = 0.01\,\text{m}$. There is a magnetic field in the z direction, only existing between $x = 0.02\,\text{m}$ and $x = 0.03\,\text{m}$, that varies as $B = 0.7y$. It will be seen from the figure, by following electron trajectories, that an inverted image is formed, with a magnification of approximately two.

4.5.1. *The transmission electron microscope*

There are several kinds of electron microscope, the most straight-forward being the *transmission electron microscope* (TEM), where the image records the transparency to electrons at different points of a thin specimen. The various components of an electron micro-scope parallel very closely those of a normal optical microscope. In place of the light source there is the electron gun. A condenser sys-tem concentrates the electrons onto the specimen to illuminate it brightly but also to restrict their divergence from the axis of the microscope, something that, just as for optical microscopes, reduces aberrations. As in the optical system an objective lens forms an image but there is the extra requirement of transforming the image into a visible form. A fluorescent screen is used that transforms the energy of the electrons into visible light, giving an image that either can be viewed directly or photographed. Figure 4.19 shows a schematic typical arrangement for a TEM.

The specimen for the TEM is in the form of a thin slice of the object being imaged; it is necessary for the electrons to be able to

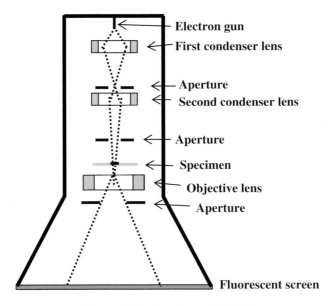

Figure 4.19   A schematic TEM.

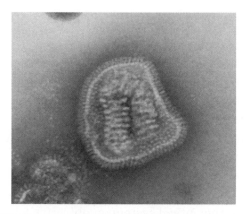

Figure 4.20   A TEM image of the influenza virus, *orthomyxoviridae* (Frederick Murphy).

pass through it. In Fig. 4.20 there is a TEM image of a cross-section of the influenza virus *orthomyxoviridae* showing clearly its protein coat and the single coil of ribonucleic acid (RNA), the genetic material that characterizes the virus. The average dimension of the virus is of order 80 nm.

4.5.2. *The scanning electron microscope*

The TEM is very similar to a normal optical microscope in that it uses an objective lens to produce an image of an illuminated specimen. The scanning electron microscope (SEM), works on a completely different principle. Condenser lenses finely focus electrons coming from the gun, as shown in Fig. 4.19, into a very fine spot of diameter between 0.4 and 5 nm. By means of varying electric or magnetic fields this spot is scanned over a rectangular area of the specimen. A basic design for an SEM is shown in Fig. 4.21. When the spot falls on a particular region of the specimen the electrons penetrate a small distance — anywhere from 100 nm to 5 μm — into the specimen interacting with all the matter it encounters. There are several different kinds of interaction of the electrons with the specimen. Some electrons are elastically scattered, which means that they interact with atoms and bounce off with no change of energy. The efficiency of this back-scattering process is very dependent on the

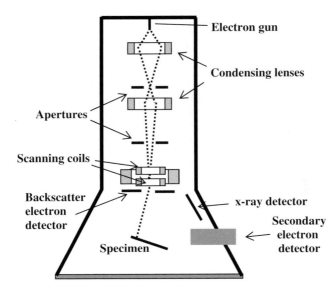

Figure 4.21   A schematic SEM.

atomic number of the atom doing the scattering, so the degree of back scattering can give information about the variation of atomic species over the scanned area. There will also be secondary electrons emitted by *inelastic* scattering in which some energy of the imping- ing electrons is given up to atomic electrons in the specimen, which then escape from the material. Finally, it is possible for electrons to lose energy by collisions and give rise to the emission of x-rays. The product of each of these interactions can be separately detected and stored in a computer memory and then used either in combination or singly to create an image; different combinations of signals can give different kinds of information about the specimen being imaged.

For the best results from SEM it is necessary for the specimen to be electrically conducting, otherwise it will acquire a negative charge at the beginning of the scanning process, thus affecting sub- sequent electrons approaching the surface, which would degrade the image. To this end, non-conducting specimens, or even some that are reasonably conducting but consist of light materials with which electrons do not interact strongly, are given a fine coat of some con- ducting material such as gold, platinum or graphite. Since specimens

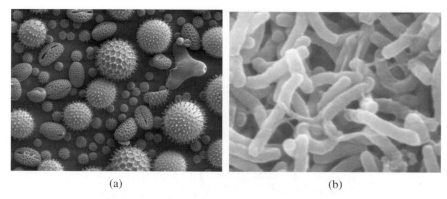

(a)   (b)

Figure 4.22   SEM images of (a) Pollen grains and (b) *Vibrio cholerae* bacteria (Dartmouth Electron Microscope Facility).

are maintained in a high vacuum some biological specimens must be specially treated both to dry them out and to preserve them under vacuum conditions.

The magnification possible with an SEM is up to 500,000 and resolution down to 1 nm, or even better can be achieved. Figure 4.22 shows two SEM images, neither near the limit of resolution of the technique but both showing a good depth of field. Figure 4.22(a) is an SEM image of pollen grains showing their detailed structures. The total width of the image is 400 μm. The image in Fig. 4.22(b) is of *vibrio cholerae* bacteria that produce cholera in humans. In this case the width of the image is only 4 μm.

4.5.3. *The scanning transmission electron microscope*

The two types of electron microscope, TEM and SEM, are the most basic and widely used and illustrate the general principles of electron microscopy. Another type is the scanning transmission electron microscope (STEM) that, as its name suggests, combines the characteristics of TEM and SEM. The electron beam is focused into a fine spot and scanned in raster fashion over a rectangular area of the specimen, as in SEM, but the specimen is thin and what is measured is the variation in the transmission of electrons through it. Figure 4.23

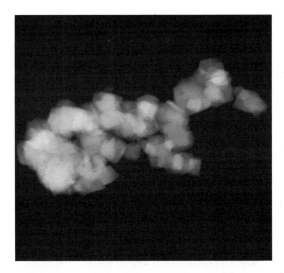

Figure 4.23   A STEM image of a very finely grained zeolite.

shows a STEM image of some finely grained zeolite. The image is some $7\,\mu$m across, showing that the smallest individual particles have a diameter of order 300 nm. At its highest resolution STEM can give images at atomic resolution.

Zeolites are a group of aluminosilicates that have a large number of important industrial uses. They can be produced in a very finely divided form and are widely used in laundry detergents, in medicine and as purifying agents. Figure 4.24 shows a model of a zeolite showing that it can act like a very fine sieve. They can also be catalysts for some industrial processes, such as the cracking of petroleum, the process by which heavier constituents of petroleum are broken down into useful lighter products.

4.5.4. *The scanning tunnelling microscope*

There is one other important type of microscope in which the passage of electrons plays an important role. The scanning tunnelling microscope (STM), as its name suggests, scans over the specimen, as does the SEM, but the type of specimen and the nature of the information gained by the imaging is quite different. The imaging is

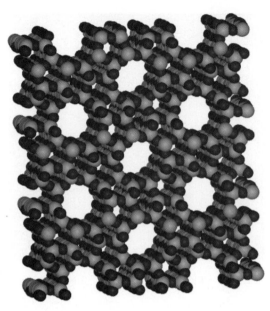

Figure 4.24   A model of a typical zeolite showing that it can act as a molecular sieve.

at an atomic level and for this reason, and also because the technical requirements of the STM involve the physics of materials that will be dealt with later, a full description of the STM will be given in Chapter 15. There it will be described in association with another imaging technique, x-ray crystallography, which also gives structural information at an atomic level.

Chapter 5

Photography and the Recording of Images

The essential components of the photographic process are, first, the creation of an image and, second, some means of permanently preserving the image. Photography is a comparatively recent invention, less than 200 years old, but knowledge relating to the essential components of photography go back a surprisingly long way.

5.1. The Origins of the Camera

The Chinese philosopher Mozi (ca. 470 BCE–ca. 391 BCE), described a *pinhole camera* some 2,400 years ago. Figure 5.1 shows this simple device for producing an image. In one face of a box is a small pinhole with the opposite face a translucent screen, for which Mozi probably used thin paper. Each point on the object is projected as a small circle on the screen; if the pinhole is very tiny then a very sharp image can be formed but only at the expense of lowering its intensity.

The small-scale pinhole camera is a precursor of the *camera obscura* ('dark room' in Latin), a term usually applied to a larger-scale device, similar in principle to the pinhole camera but housed in a room or tent, for viewing the surrounding landscape. In the tenth century the Muslim scientist Ibn al-Haytham (965–c.1040), generally known as Alhazan, who established the fact that light travelled in straight lines, observed the Sun during an eclipse with a camera obscura and noted how it took on a crescent form as the eclipse progressed. In 1490 Leonardo Da Vinci (1452–1519) described the

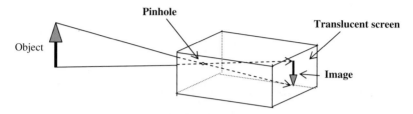

Figure 5.1   A simple pinhole camera.

construction of a camera obscura; it was a device used by many well-known artists who traced the image and then used the tracing as a template for their painting. A problem with the camera obscura was that it produced a faint and also an inverted image. In the sixteenth century a lens was used in place of the pinhole to create a much sharper and brighter image and then the introduction of a mirror to project the image onto a horizontal surface also made it possible to see a non-inverted image. From the nineteenth century onwards many camera obscuras were constructed in country houses and in museums for both entertainment and education. A typical example is in the National Museum of Photography, Film and Television in Bradford, West Yorkshire — a device that can be rotated to give a detailed view of the surrounding cityscape and the activity within it.

5.2. Recording and Storing Monochrome Images

The first steps in recording and storing images were the result of work by early alchemists who discovered that some chemicals changed colour when exposed to light. The basis of many early image storage systems was silver nitrate, $AgNO_3$, discovered by Albertus Magnus (1193–1280). When this is exposed to light it decomposes to give metallic silver that, in fine-grain form, appears to be black. Subsequently it was noticed that other chemicals also changed colour when exposed to light. In the 1790s Thomas Wedgewood (1771–1805), the son of the potter Josiah Wedgewood, was creating pictures by exposing to sunlight opaque objects placed on paper soaked in silver nitrate. The paper became uniformly black if the paper was kept in the light so Wedgewood stored his images in a darkened room where they could only be viewed in dim candlelight. Wedgewood is sometimes called 'the father of photography' although there are certainly others more deserving of that accolade.

5.2.1. *Joseph Nicéphore Niépce*

A strong candidate for the title of 'the father of photography' must be the French inventor Joseph Nicéphore Niépce (1765–1833), whose name is little known to the general public. Niépce produced the first permanent photographic image. His process began by coating a metal plate with a solution of bitumen, a viscous black oily substance, in lavender oil. It was then exposed to an image produced by a camera obscura for about eight hours. The effect of the exposure was to bleach the bitumen to a pale grey colour and also to harden it in the places where light had fallen. The plate was then washed in lavender oil that removed everything except the hardened bitumen from the plate. A permanent positive image was formed. The dark grey of the metal substrate provided the dark parts of the image and the bleached bitumen gave the paler regions. Intermediate intensity of exposure gave partial bleaching and partial hardening so that not all the bitumen was removed. The earliest image produced by this process, which Niépce called *heliography* (sun writing), dating from 1826, is shown in Fig. 5.2.

Figure 5.2   View from a window at Le Gras.

The long exposures required by this process severely limited its usefulness. Niépce continued his attempts to find a more practical process, this time in collaboration with a compatriot, Louis Daguerre (1787–1851), who was well known as a designer in the theatre world and was also a chemist. The basis on which Niépce and Daguerre built their work was the discovery by the German scientist Johann Heinrich Schultz (1684–1744) that a mixture of silver nitrate and chalk blackened when exposed to light. However, before the pair had achieved anything of note Niépce died and Daguerre carried on the work alone.

5.2.2. *Daguerreotypes*

By 1837 Daguerre had created a process that gave detailed permanent images on a timescale of a few minutes — the *Daguerreotype*. In this process a silver-coated copper plate was exposed to iodine vapour that converted the silver into silver iodide. This plate was then exposed to the image for several minutes; the greater the exposure at a point on the plate the more silver was produced. Next the plate was exposed to mercury vapour that converted the deposited silver into silver amalgam — a shiny compound of silver and mercury. Finally the plate was washed in salt water, which removed everything from the plate except the amalgam that formed a silvery deposit on the plate, the density of which was proportional to the exposure to light. The image was permanent but was laterally inverted (right to left and *vice versa*), although this could be corrected by laterally inverting the image with a mirror before it fell on the plate. The Daguerrotype had to be viewed obliquely in light reflected from the plate, so that where the amalgam was thickest, i.e. in the brightest part of the image, the plate appeared brightest. If viewed directly the image was seen as a negative. An early photograph, of a Paris street scene, taken by Daguerre in 1839, is shown in Fig. 5.3. Because the exposure was several minutes the moving traffic is not seen. However, a man appears in the bottom left of the picture; he was having his shoes shined and stayed still for the requisite period. This unknown man has the distinction of being the first ever to be photographed.

Figure 5.3   A Paris street scene taken by Daguerre.

5.2.3. *William Henry Fox Talbot*

In parallel with Daguerre's work a British inventor, William Henry Fox Talbot (1800–1877), was also working on a process of producing permanent images; he had achieved this a few years earlier than Daguerre but had not made his work public. Eventually, in 1841, he patented a process for producing what he called *calotypes* ('kallos' in Greek means 'beautiful') that introduced an important new principle. The first stage produced a *negative image*, one in which light and dark were reversed so that the brightest parts of the object scene were darkest in the negative and *vice versa*. From this negative as many positive images as were required could be produced. This was a considerable advance on the Daguerrotype that could only produce a single final image.

The calotype process started with a high-quality smooth sheet of paper that was soaked in a solution of silver nitrate and then partially dried. It was then immersed in a solution of potassium iodide, which produced silver iodide in the paper, and then rinsed and dried. This process had to be carried out in weak candlelight since the paper being produced was photosensitive. Not long before the photograph was to be taken the paper was soaked in a freshly prepared solution of *gallo-nitrate of silver*, an equal mixture of silver nitrate and gallic acid ($C_6H_2(OH)_3COOH$), and then dried again. This was loaded into the camera and the photograph taken; exposure times varied from about 2 seconds to several minutes, depending on the brightness of the scene. When the paper was taken from the camera no image was visible but when immersed in a fresh solution of gallo-nitrate the latent image in the paper gradually developed. When the photographer judged that the image was of suitable density the paper was immersed in a *fixer*, a solution that removed all remaining photosensitive material from the paper. This could be potassium bromide or sodium thiosulphate ($Na_2S_2O_3$), known to modern photographers as *hypo*.

The paper for producing the positive images was first soaked in common salt (sodium chloride), and then brushed on one side with silver nitrate solution to produce photosensitive silver chloride. The negative was placed in contact with this print paper and then exposed to bright sunlight for fifteen minutes or so, which produced an image on the print paper. This was finally fixed with hypo, rinsed in clean water and dried. An early calotype image is shown in Fig. 5.4.

Figure 5.4   A calotype image of Thomas Duncan taken in 1844 (National Gallery of Scotland).

In the mid-1840s there were two practical ways of producing permanent images. Fox Talbot's process was by far the more practical in that large numbers of positives could be produced, but daguerreotypes gave an image of far better quality. What was needed was a combination of the flexibility of calotypes with the image quality of daguerreotypes.

A considerable, but little recognized, contributor to the development of photography in this period was John Herschel (1792–1871), the son of William Herschel, the astronomer who discovered the planet Uranus. John Herschel was a man of many talents — a mathematician, astronomer, chemist, botanist and inventor in the field of photography. He invented the process for making blueprints and it was he that discovered that hypo dissolved both silver chloride and silver iodide and hence could be used as an agent for fixing images. He also introduced the name *photography* for this way of producing images and the terms *negative* and *positive*.

5.2.4. *From the wet collodion process to modern film*

In 1851 the British sculptor Frederick Scott Archer (1813–1857) invented what is called the *wet collodion* process for photography, which quickly became the dominant process. Collodion is a rather sticky syrupy solution of cellulose nitrate in ether and alcohol. A glass plate was first uniformly covered with collodion containing a halide salt (chloride, bromide or iodide). Then, in a darkroom, this plate was immersed in a solution of silver nitrate for a few minutes, which produced one of the silver halides suspended in the collodion. This plate was loaded into a slide holder and inserted into the camera; according to the lighting conditions exposures could vary from a few seconds to several minutes. The plate was developed in a developer consisting of ferrous sulphate, acetic acid and alcohol and then fixed either in hypo or potassium cyanide. With sufficient care this produced a negative of very high quality; Fig. 5.5 shows a print produced from such a negative.

The wet-plate process gave photographs of excellent quality but since the plates had to be prepared immediately before the

Figure 5.5   A wet-collodion photograph of Theodore Roosevelt (Library of Congress).

photograph was taken it was difficult to take photographs outside a studio or some other place equipped to carry out the preparation. For taking photographs away from such facilities, such as for those taken of scenes from the American Civil War, photographers used portable studios or tents to prepare their plates. In 1871 an English doctor, Richard Maddox (1816–1902), invented the *dry plate* in which the photosensitive chemicals were contained in a layer of gelatine on a lightweight glass plate. These could be stored in light-tight boxes and used when required. After some improvements to increase their resistance to damage, in 1878 the American inventor, George Eastman (1854–1932), the founder of the Eastman Kodak Company, perfected a way of producing dry plates on a commercial scale. By 1889 the Kodak Eastman Company was also producing *roll film*, where the impregnated gelatine was coated on celluloid, together with an easy-to-use camera. This development moved photography from being a specialist activity into the mass market.

5.3. The Beginning of Colour Photography

The first photography was all in monochrome, generally black and white but sometimes the final prints appeared in sepia or even a slightly purple hue. In 1861 the Scottish physicist James Clerk Maxwell (1831–1879) devised an arrangement for producing a full-colour image. The basis of his method was that all colours with any degree of *saturation* (mixture with white) can be reproduced by mixing three *primary colours* — red, green and blue. This related to the way in which colour is seen by the eye (§1.2). Maxwell took three black-and-white negative photographs of a scene, all from precisely the same position, one through a red filter, the second through a green filter and the third through a blue filter. Positives of these were produced on glass slides; the transparency of the red positive would be greatest where that part of the scene had the greatest red component and similarly for the green and blue positives. Then the three positives were projected simultaneously to give precisely overlapped images with the red positive projected through a red filter and similarly for the other two positives. The result on the screen was a full-colour picture — the first one ever seen in public — of a tartan ribbon. It was produced by Maxwell at a Royal Institution lecture in 1861 and is shown in Fig. 5.6. However, it was a projected image — not a photograph.

5.3.1. *Louis Ducos du Hauron*

Louis Ducos du Hauron (1837–1920), a French scientist who in 1869 wrote an influential book *Les Couleurs en Photographie*, produced

Figure 5.6   Maxwell's full-colour projection of a tartan ribbon in 1861.

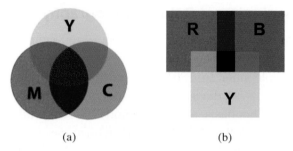

(a)   (b)

Figure 5.7   Subtractive colour, produced by the use of (a) yellow (Y), cyan (C) and magenta (M) pigments and (b) yellow (Y), red (R) and blue (B) pigments.

the first true full-colour picture. The method he used depended on the concept of *subtractive colour*, the principle of which is illustrated in Fig. 5.7.

In Fig. 5.7(a) the yellow pigment looks the way it does because it has absorbed the blue part of the white light that fell upon it and the reflected light, white minus blue, looks yellow. The cyan (green-blue) pigment looks as it does because it absorbs the red end of the spectrum while the magenta (red-purple) absorbs the yellow-green part of the spectrum. Now we consider what happens when the yellow and cyan pigments are both present. The yellow pigment absorbs the blue part of the spectrum and the cyan pigment absorbs the reddish end. The residual wavelengths that are reflected are dominantly in the green part of the spectrum and so the overlapped pigments look green. The other overlapped regions can be similarly explained. Where all three pigments overlap virtually all spectral wavelengths are absorbed and the net appearance is black. Most modern colour printers used in conjunction with home computers use admixtures of yellow, magenta and cyan pigments to produce the full range of colour effects. Similarly, as seen in Fig. 5.7(b), yellow, red and blue can be used as the set of three basic subtractive colours.

One of the problems faced by Ducos du Hauron when he first considered the problem of producing a coloured photograph was that the photographic plates available were insensitive to red light. However, in 1873 a German chemist and photographer, Hermann

Wilhelm Vogel (1834–1895), discovered that adding the yellowish dye corallin, derived from coal tar, to the collodion that coated the plate increased the sensitivity to the red-yellow end of the spectrum; other dyes could give other changes of colour sensitivity and this discovery enabled Ducos du Hauron to implement the theories he presented in his 1869 book.

With plates dyed to be sensitive in the appropriate colours, Ducos du Hauron made black-and-white glass negatives of a scene taken through orange, green and violet filters. To produce a print of the original scene he prepared three sheets of paper each coated with gelatine containing a different pigment — carmine (crimson), Prussian blue and orpiment (yellow to yellow-orange). These were rendered photosensitive by the addition of ammonium dichromate $((NH_4)_2Cr_2O_7)$ dissolved in alcohol. These paper sheets were then placed in contact with the negatives and exposed to light — the yellow pigment with the violet-filtered plate, the red pigment with the green-filtered plate and the blue pigment with the orange-filtered plate. To consider the effect of this we consider the violet-filtered negative. The violet filter would be opaque to light in the yellow region of the spectrum so a yellow region of the scene would be clear on the negative. This means that the paper sheet with yellow pigment would be most exposed to the yellow parts of the scene. The effect of the dichromate on the gelatine is that when it is exposed to light it hardens. By washing the final plate the unhardened gelatine is removed leaving coloured gelatine approximately in proportion to the transparency of the negative at each point. The gelatine from each sheet was then transferred to a separate glass plate by pressing the plate against it. Each glass plate was then pressed, in exact registry on the same piece of paper to give the final print. The relationship between the colour of the object and the print is given in Table 5.1.

Clearly the process was demanding and time consuming but could give good results. Figure 5.8 shows a print of the city of Agen, in the Aquitaine Region of France, produced by Ducos du Hauron in 1877.

Table 5.1   The Ducos du Hauron process for producing a colour print. Each entry gives at the top the density of the negative and at the bottom the contribution to the print.

Filter and gelatine colour	Object colour			
	Red	Yellow	Green	Blue
Orange filter	Black	Black	Moderate	Clear
Blue gelatine	None	None	Some	Strong
Green filter	Clear	Moderate	Black	Moderate
Red gelatine	Strong	Some	None	Some
Violet filter	Clear	Clear	Moderate	Black
Yellow gelatine	Strong	Strong	Some	None
Net result on print	Red with orange tinge	Yellow with orange tinge	Green	Blue with purple tinge

Figure 5.8   A photograph of Agen taken by Ducos du Hauron in 1877.

5.3.2. *The Lippmann process*

A completely different process for producing a true-colour photograph was described by the French physicist Gabriel Lippmann (1845–1921; Nobel Prize for Physics, 1908) in 1897. This process depends on the wave nature of light and is illustrated in Fig. 5.9. It

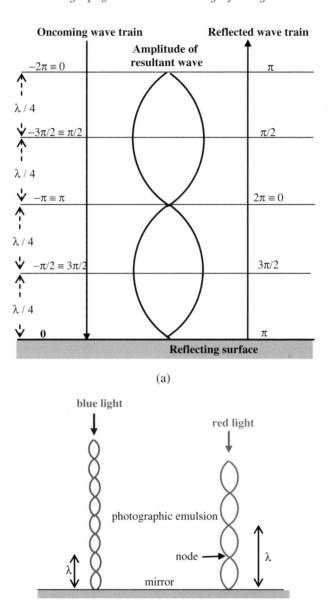

Figure 5.9 The interference pattern set up when coherent monochromatic light is reflected. (a) Phases of incident and reflected wave train relative to that of the incident light at the mirror surface. (b) Representation of the net amplitude at different distances from the mirror for both blue and red light.

depends upon the fact that when light is reflected from a mirror its phase is changed by π, which means that at the surface the incident and reflected waves cancel each other out giving zero intensity. Normal light consists of a series of short wave trains, typically about ten wavelengths or so in length, which means that interference between the incident and reflected light can only occur in a region within a few wavelengths of the mirror. The incident light at a distance of $\lambda/4$ from the mirror will always be $\pi/2$ in phase behind the oncoming wave at the mirror surface and the phase delays at distances of other multiples of $\lambda/4$ are also shown. For the reflected wave at a distance of $\lambda/4$ the phase will be $\pi/2$ ahead of the reflected wave at the mirror; the phases shown for the reflected wave at various distances from the mirror are shown relative to the phase of the incident wave at the mirror surface. Wherever the phases are the same the resultant will have maximum amplitude and wherever they differ by π there will be zero amplitude (a node). The amplitude of the resultant wave motion is shown in the figure.

The Lippmann process uses a glass plate coated with a very thin emulsion, with thickness just a few wavelengths of light, containing extremely fine photosensitive grains. For the reflecting mirror a mercury surface is used on which the plate is placed with the emulsion side in contact with the mercury to give the required coherence within the emulsion. If monochromatic light were projected onto the plate then, wherever the amplitude was greatest, as shown in Fig. 5.9(a), the plate would be blackest and wherever a zero amplitude occurred, at a node, the plate would be clear. If the light contained a spread of wavelengths then each of them would have a resultant as shown in Fig. 5.9(b) and the overall blackening pattern would be a sum of the blackening from all the wavelengths present. When white light is projected onto the developed plate the silver grains scatter the light and that which is scattered in the direction of the reflected light interferes in such a way that it reproduces the wavelength distribution that came from the object, so the correct colour is seen in the image.

An example of an early Lippman colour picture is shown in Fig. 5.10.

Figure 5.10   An early Lippmann colour photograph.

5.4. Modern Colour Photography

So far we have considered three methods for forming a coloured image — that of Maxwell, which is a projected image using an additive process with three primary colours, that of Ducos du Hauron depending on subtractive colour produced by overlapping pigments and that of Lippmann that uses the interference of light. What is clearly desirable is to have a process similar to that of black-and-white photography, one that gives a negative that can subsequently be used to produce any number of positive images.

5.4.1. *The autochrome process*

The brothers Lumière — Auguste (1862–1954) and Louis (1864–1948) — who were eminent French motion film-makers, invented the first commercial colour photography process in 1907. The basis of the autochrome process was to coat a glass plate with a single layer of transparent potato starch grains, between 5 and $10\,\mu m$ in diameter, some dyed red, some green and some blue — although other combinations of colour are also possible. The spaces between the grains were filled with lampblack so they would not transmit light. The grains were then protected with a layer of transparent shellac on top of which the photographic emulsion containing the photosensitive silver halide was deposited.

The image was projected through the glass plate so that the light passed through the starch grains before reaching the emulsion. Thus if red light falls on a red grain then the light passes through it and a silver deposit is produced below it. However, if the light falling on the red grain is green or blue then it will be blocked and no silver, or little silver, will be produced below it. The plate is then developed but it is not fixed as is usual for black-and-white photography. Instead the plate is immersed in a solution that dissolves the silver but leaves the silver halide unaffected. Next the plate is exposed to light, so producing silver where the silver halide remained; the original negative image on the plate has been reversed. Where red light fell on the plate the red grains are now exposed so that if the plate is viewed in transmitted light the colour red is seen. In this way, when the plate is placed in a viewing frame, what is seen is a full-colour picture, an example of which is shown in Fig. 5.11.

Although the autochrome process gave a good coloured image, it does not give what is required — a coloured negative from which many positive prints can be produced.

Figure 5.11   An Autochrome photograph of Algerian soldiers taken in 1917.

5.4.2. *The modern era of colour photography*

The next significant development in commercial colour photography was made by the American Kodak Company in the 1930s and was the invention of an unlikely duo — Leopold Mannes (1899–1964) and Leopold Godowsky (1870–1938), two professional musicians. In this subtractive-colour process, known as *Kodachrome* and illustrated in Fig. 5.12, the colour film was covered by three layers of emulsion respectively sensitized to the three primary colours — red, green and blue. Silver deposits are formed in the three layers as shown in Fig. 5.12(a). When the film is developed the negative images in each layer are reversed as described for the autochrome process so that the silver is removed and silver halide remains, which subsequently converts to silver. However, while the reversal is happening another process is taking place. The developer becomes oxidized and combines with *colour coupler* compounds in the developer to produce different dyes in the three layers — cyan, magenta and yellow in the red-, green- and blue-sensitized layers respectively. The situation is now as shown in Fig. 5.12(b) with the mixtures of silver and dyes in the three layers. Finally all the silver is chemically removed to give the situation illustrated in Fig. 5.12(c). As an example of the effect of this process, in the red part of the scene white light passes through magenta pigment, which removes green, and through yellow pigment,

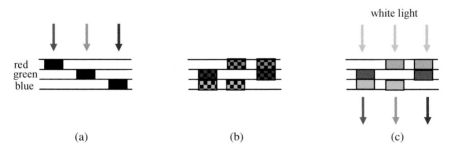

Figure 5.12   Stages in the Kodachrome process. (a) Negative black-and-white images for the three colours. (b) Reversal of the negatives and production of colour subtractive dyes. (c) The removal of residual silver grains.

which removes blue, thus transmitting only red light. The stages of this process are illustrated schematically in Fig. 5.12 in terms of red, green and blue light falling on the film; if other colours fall on the film then the red, green and blue components in the colour will give effects in each of the three layers of emulsion in proportion to their intensities. The final product, seen in transmission or projection as a slide, shows the colours of the original incident light on the film.

The developing process was extremely complicated and could not be performed by amateur photographers. In 1936 the German company Agfa produced a colour film based on the same general principle but with the colour couplers incorporated within the original layers in the emulsion, which gave a much simpler developing process.

Although the process as described here is for coloured slides it is possible to obtain colour negatives from the slides that, used with specially coated paper, can produce colour prints. In 1942 Eastman Kodak introduced *Kodacolor*, a colour negative film produced specifically to give colour prints. However, by the end of the twentieth century the use of film for photography was much reduced and Kodak ceased to provide a processing service to produce colour prints in 2009.

5.5. The Basic Construction of a Camera

The use of a lens in a camera obscura to produce an image that was simultaneously sharp and bright was mentioned in §5.1. The lenses of modern cameras are extremely complex in construction. As explained in §4.3, a simple convex lens does not give a satisfactory image because of *dispersion*, the variation of refractive index of the lens material with wavelength, giving rise to chromatic aberration. In expensive cameras achromatic compound lenses can consist of several optical elements.

Another feature of high-quality lenses used for cameras is the process of *lens blooming*. When a ray of light hits an interface between two materials of different refractive indices some of the light is reflected, leading to a loss of transmitted light — which is undesirable in a camera lens. We noted in describing the Lippmann process that there was a phase change of π when light was reflected from a mirror. Actually, this phase change only occurs when the light from one medium falls on another of higher refractive index, i.e. when light goes from air to glass but not when it goes from glass to air. Lens blooming consists of coating the outside surface of the lens with a layer of a transparent material the refractive index of which is less than that of the glass of the lens. For the blooming process to work the thickness of the layer should be a quarter-wavelength of the light corresponding to the refractive index of the layer material; because of dispersion, this cannot be achieved simultaneously for all the wavelengths in white light so the thickness is chosen for a green wavelength in the middle of the visible spectrum. The behaviour of the light falling on the lens is illustrated in Fig. 5.13. Light reflected at the air medium interface has a phase change of π because the medium has a higher refractive index than air. The light reflected at the medium-glass interface also undergoes a phase change of π on reflection but by the time it reaches the medium-air interface has an additional phase change of $\pi/2 + \pi/2 = \pi$ giving a total phase change of 2π, equivalent in phase terms to zero. This is π out of phase with that directly reflected at the air-medium interface so giving destructive interference that reduces the total amount of light reflected and hence higher transmission. Since this process only works precisely for a particular green wavelength and is less efficient the further from that wavelength, such reflection as does occur is mostly from the blue and red ends of the spectrum, giving bloomed lenses a purplish hue.

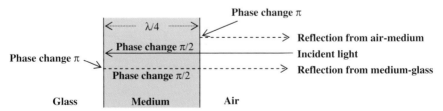

Figure 5.13   Lens blooming. There is a phase change of π at each reflection and $\pi/2$ for each passage through the Medium.

Another desirable feature of a camera lens is that it should have a large diameter and so admit a great deal of light, combined with a small focal length so that the camera can be of a convenient size and so that the image is small and therefore bright. The ratio of the focal length of the lens to its diameter is called its *f-number* and a good camera might typically have an f-number of 1.4. However, with a small f-number there is a reduced *depth of focus*, meaning that the range of object distances for which the image is reasonably sharp is restricted. For some photographic purposes this can be a desired feature; the photographer may wish to have a sharply focused foreground and an out-of-focus background. Alternatively if a large depth of focus is required then it is necessary to have a smaller f-number and this is done by *stopping down* the diameter of the lens using an *iris diaphragm*, so-called because it behaves like the iris of the eye in adjusting the diameter of the pupil (Fig. 1.2). The dependence of the depth of focus on the f-number is illustrated in Fig. 5.14 that shows what happens when images of two point objects at different distances from a lens are produced with different f-numbers. The image of A is at A' on the film while the image of B is at B', in front of the film. In Fig. 5.14(a) the extreme rays passing

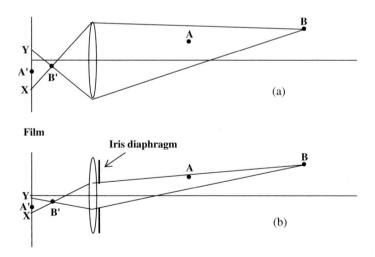

Figure 5.14   The effect of stopping down a lens on the depth of focus.

through the lens strike the film at X and Y so the image of the point B on the film is a circle of diameter XY. Where the diameter of the lens is restricted by an iris diaphragm, as in Fig. 5.14(b) the image of point B on the film is a circle of much smaller diameter. This means that reasonable focusing can be obtained over a larger range of distances.

Stopping down a lens reduces the intensity of light on the film. To get the best quality of photograph it is necessary to expose the film neither too much, giving a dark image with reduced contrast, nor too little, giving a faint image. There are several ways of ensuring a correct exposure. Films are available of different sensitivity, some very sensitive that are well suited to poorly illuminated subjects. Another control is the shutter speed of the camera, i.e. the length of time for which the film is exposed. If the object is moving, say a flying bird, then clearly a very short exposure, say 1/1,000th of a second, is desirable. A skilled photographer can manipulate the controls at his disposal to obtain the results he desires. For example, if he wishes to photograph wildlife, which may be mobile, but also for artistic reasons wants his subject to be in focus but background to be much less focused then he can use a very sensitive film, large f-number and a very short exposure time. Alternatively, he may be photographing landscapes, which do not move so that exposures can be longer, and where light conditions allow a less sensitive film to be used, which may give a finer film grain and hence give a photograph with more detail. For the majority of users, who just take family and holiday photographs, manufacturers have created cameras with automatic focusing, aperture and shutter timing that give tolerably good results over a wide range of conditions.

5.6. Digital Cameras

In 1969 the Canadian physicist William Boyle (b. 1924) and American physicist George E. Smith (b. 1930), working at the Bell Research Laboratory in America invented the *charge-coupled device* (CCD), for the invention of which they jointly received the Nobel Prize for Physics in 2009. This electronic device has several applications, one of which is to record and store an image. It basically consists of a silicon crystal, typically about 1.5 cm square, on which have been fabricated a two dimensional array of light-sensitive elements, which correspond to the *pixels* (picture elements) of the final image that is to be formed. The first CCDs manufactured were comparatively crude devices with up to 100×100 elements but the number of elements in modern CCDs, each about 0.03 mm in side, is normally several million and they are capable of giving a very high resolution image.

The light falling on each element is converted into an electronic charge that is proportional to the intensity of the light falling on it. The charge on each element is then transferred along a sequence of elements (hence *charge-coupled*) and when it reaches the end it is converted into a voltage that can then be stored in the digital camera's memory. This record can subsequently be downloaded into a computer or printed on a digital printer with each CCD element giving a different pixel of the final image.

A CCD detects about 70% of the light falling on it, making it much more sensitive than film that utilizes only 2% of the incident light. However, it cannot distinguish light of different colours so to use a CCD for colour photography it is necessary to introduce coloured filters. An array of red, green and blue filters is placed over the CCD elements and the signal from each element must be represented in the same colour at the corresponding pixel of the final image. The usual filter used is a *Bayer filter*, designed in 1976 by Bruce E. Bayer of the Eastman Kodak Company and illustrated in Fig. 5.15. This has two green elements for every red or blue one, which simulates the overall sensitivity of the eye that is most sensitive in the green region of the spectrum.

Figure 5.15   A section of a Bayer filter array.

Chapter 6

Detecting and Imaging
with Infrared Radiation

There is a saying that 'War is the mother of invention', which is actually a corruption of an original saying of Plato that 'Necessity is the mother of invention'. Many techniques for imaging and detection by the use of non-visible radiation were first developed with military applications in mind but have turned out to have great utility in civil life. Thus radar, which was initially motivated by the requirement of detecting the approach of enemy aircraft, is now a key element in the efficient and safe running of airports and also safety at sea. However, we shall begin our discussion of detection and imaging with non-visible radiation by the use of electromagnetic radiation of wavelength moderately longer than that of visible light — *infrared radiation*, sometimes called heat radiation. One early example of this was the German installation of devices using infrared radiation to detect ships moving through the English Channel during the Second World War.

6.1. The Radiation from Hot Bodies

All objects at a finite temperature emit electromagnetic radiation but the quantity, type and distribution of the wavelengths they emit is highly dependent on their temperature. We are all familiar with the way that materials emit radiation when they are heated. An iron object at the temperature of heated water, as in a domestic radiator, emits heat but is quite invisible in a darkened room. An iron used for pressing clothes is at a considerably higher temperature than the radiator, and is certainly a more efficient radiator of heat, but is still invisible in a darkened room. In days past, when coal fires were more common than they are now, it was usual to use a poker to disturb the burning coals to invigorate the fire. A poker left in the fire would emit a dull red glow; it would produce visible light. When a horse is shoed the blacksmith heats the iron shoe to a high temperature by bellows so that it glows brightly; the iron can then be hammered to the correct shape for the horse's hoof. To see iron at an even higher temperature one would need to visit a steelworks in which iron is molten and glows with a brilliant reddish-orange light that is so bright that to look into a steel furnace requires the use of protective goggles. To see the output from even higher temperatures one can look at the light source within a filament lamp. For scientific purposes the temperature scale used is the *Absolute* or *Kelvin* scale (unit *kelvin*, symbol K). Temperature is a measure of the energy of motion of the atoms in a substance and the temperature on the Kelvin scale is proportional to that energy. At 0 K, the atoms are at rest.[1] The scale has the same increments as the Celsius scale; the melting point of ice is 273 K and the boiling point of water 373 K. In a vacuum lamp the temperature of the tungsten filament could be between 1,800 and 2,700 K and with increasing temperature the emitted light gets both brighter and moves from reddish towards a whitish appearance. With a gas-filled filament lamp, which can be run up to 3,200 K, the light is clearly white and, when broken up into its spectral components with a prism, shows a large component of wavelengths at the blue end of the spectrum.

The German physicist Max Planck (1858–1947; Nobel Prize for Physics, 1918) gave the theory for the distribution of intensity with wavelength for emission from a body at any given temperature. This theory marked an important landmark in physics because it contained the idea that electromagnetic energy could only exist in discrete units related to its frequency — the beginnings of quantum

---

[1] According to quantum theory, even at 0 K there is energy of motion called *zero point energy* — but it is tiny and we can disregard it.

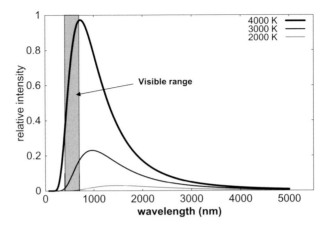

Figure 6.1 The radiation curves for 4,000, 3,000 and 2,000 K.

theory that was later used by Einstein to explain the photoelectric effect. The relative distributions of intensity with wavelength are shown for temperatures 2,000, 3,000 and 4,000 K in Fig. 6.1 and several features are evident. The intensity is the measure of energy per unit time per unit increment of wavelength emitted by a unit area of the body. From this it follows that the total energy over all wavelengths emitted per unit area per unit time is the area under the curve and this will be seen to be very heavily temperature dependent. Planck's theory shows that it is proportional to T^4 where T is the absolute temperature. Many hot bodies emit radiation very similarly to *black bodies*, which are ideal radiators emitting the maximum possible radiation from their surfaces. The total energy emitted per unit area per unit time by a black body is

$$I = \sigma T^4, \tag{6.1}$$

in which σ is Stefan's constant, $5.67 \times 10^{-8}\,\mathrm{W\,m^{-2}K^{-4}}$.

Another feature that is evident from Fig. 6.1 is that the peak of the curve moves towards smaller wavelengths with increasing temperature. Wien's displacement law gives the relationship between the wavelength of maximum intensity, λ_{max}, and absolute temperature,

T as

$$\lambda_{max}T = 2.9 \times 10^{-3} \text{ m K}. \tag{6.2}$$

The visible range of wavelengths is indicated in the figure, from which it will be seen that such light as is emitted at 2,000 K is all at the red end of the spectrum. At 4,000 K more of the blue end of the spectrum is present so that the light has a whiter appearance. Another obvious feature, even at 4,000 K, is that most of the energy is produced at longer wavelengths, as heat, and comparatively little as light, hence explaining the inefficiency of filament bulbs and the drive to get them replaced with more efficient long-life bulbs that produce light by fluorescence.

It is impossible to display the radiation curves for much lower temperatures in Fig. 6.1 at the same scale because they would be too close to the x-axis. Figure 6.2 shows the radiation curves for 500, 400 and 300 K; to get a vertical scale comparable to Fig. 6.1 the y-axis values must be divided by 33,000. What is clear from this figure is that there is no effective radiation of visible light and it is all in the longer-wavelength infrared to heat region of the electromagnetic spectrum. To give an idea of what kind of bodies radiate in this range, the human body is at a temperature of about 311 K, boiling

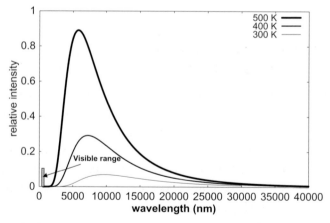

Figure 6.2   Radiation curves for temperatures at which only heat radiation is emitted.

water is at $373\,\text{K}$ and the optimum temperature for an iron to press cotton fabric is about $480\,\text{K}$. These are all temperatures met with in everyday life and are the sources of easily detectable amounts of heat radiation.

We can now see the requirements and problems of detecting infrared radiation. Very hot bodies radiate considerable amounts of infrared radiation, but also much visible light so they are most easily detected or imaged by the latter. Cool bodies emit no discernable visible light, so cannot be detected or imaged using that radiation. However, although they emit infrared radiation they are comparatively poor emitters so it is necessary to have sensitive detectors to detect them and special optics to image them.

6.2. The Detection of Infrared Radiation

There are several different kinds of infrared detector but they can be divided into two basic types. The first type, *thermal detectors*, exploits the heating effect of infrared radiation when a body absorbs it. In that sense we are all infrared detectors since we know when we stand near a hot object — a working oven, for example — just by the radiation that we absorb from it. The second type, *photonic detectors*, uses materials that directly convert the energy of infrared photons into an electric current.

6.2.1. *The effectiveness of infrared and heat detectors*

To compare the efficacy of several different methods of making any particular measurement it is necessary to find some measure of their sensitivity. In systems for measuring infrared radiation the normal output is a voltage generated in a resistor the magnitude of which indicates the power of the radiation. The *responsivity* of the system, R_V, is defined as the voltage developed across the resistor, V, per radiant power Q, incident on the infrared detector, or

$$R_V = V/Q \qquad (6.3)$$

While responsivity is a useful measure of the sensitivity of a detecting system it takes no account of an inherent problem with any system attempting to measure very weak signals, which is the presence of random noise. A system with a high responsivity that also has a high noise level may give a less precise measurement than one with a smaller responsivity but much lower noise level. The manufacturers and suppliers of infrared and heat detection equipment quote a quantity called *Noise Equivalent Power* (NEP) to describe its effectiveness but, unfortunately, they do not all agree on how to define it. We now give one measure, which will be quoted in relation to various types of measuring systems.

The noise in a measurement could be caused by fluctuations in the number of photons received by the device due to the signal; if the number of photons being received per unit time is n then the expected (mean) number received in time t is nt. If the actual number received in a particular time interval t is M then the deviation from the mean is $M - nt$ and the square deviation from the mean is $(M - nt)^2$. Averaged over many time intervals t there would be a *mean square deviation* and the square root of this is the *standard deviation*, a measure of the extent to which the number of photons in a time interval t fluctuated. Theory shows that the standard deviation is $\sqrt{nt}$. The ratio of signal, S, to noise, N, can be represented by

$$\frac{\text{Signal}}{\text{Noise}} = \frac{nt}{\sqrt{nt}} = \sqrt{nt}. \qquad (6.4)$$

Another, and potentially more important, source of noise, *Johnson–Nyquist noise*, is inherent in the measuring equipment and is due to random fluctuations of the motions of charge carriers in the equipment. It is *white noise* with a wide range of frequency components and the power it generates is roughly constant per unit frequency range.

When a detector is making a measurement what it gives is the integrated signal it receives over some period of time τ, which might be an inherent characteristic of the detector or something that can be controlled. The fractional fluctuation in the integrated intensity due to one frequency component of the signal will fall off with the number of oscillations there are of that component within the period τ. The assumption used here is that if τ is, say, 0.01 s, then any frequency in the noise greater than $1/\tau$ —100 Hz^2 in the case we take here — will not greatly affect the reading. If the range of noise frequencies (*bandwidth*) that could influence the measurement is Δf then the total power of the noise in that range of frequencies is proportional to Δf. Now this power is equivalent to a random voltage V_r; in any given electric circuit voltage is proportional to the square root of the power generated from which we can write

$$V_r = N\sqrt{\Delta f} \tag{6.5}$$

where N is a constant for the equipment and is a measure of its effectiveness. The smaller is N the less noise there is for a given range of measurement frequencies. The definition we use here for NEP is

$$\text{NEP} = \frac{N}{R_V} = \frac{V_r}{\sqrt{\Delta f}}\frac{Q}{V} \tag{6.6}$$

from which it will be seen that it is measured in the units $\text{W Hz}^{-\frac{1}{2}}$. The smaller is the NEP the better is the detector in terms of having a small ratio of noise to signal.

We now consider a detector for which NEP $= 10^{-10}$ $\text{WHz}^{-\frac{1}{2}}$, $\Delta f = 100$ Hz and the power of the received radiation is $Q = 10^{-6}$ W. Substituting these values in (Eq. 6.6) we find the signal to noise ratio

$$\frac{\text{Signal}}{\text{Noise}} = \frac{V}{V_r} = \frac{Q}{\sqrt{\Delta f} \times \text{NEP}} = 10^3. \tag{6.7}$$

For a detector responsivity of $10\,\text{VW}^{-1}$ the signal would be 10^{-5} V and the noise 10^{-8} V.

If the power of the received radiation is very small, such as when recording the radiation from astronomical objects, then the incoming signal can be recorded over a long period of time, thus reducing Δf and increasing the signal to noise ratio — although the signal will still be small and require careful measurement.

---

[2]The unit 1 hertz (Hz) is 1 cycle per second and has dimension time^{-1}.

6.2.2. *Thermocouples and thermopiles*

Thermal detection was the basis of the discovery of infrared radiation. The British astronomer William Herschel (1738–1822) carried out an experiment with a spectrum produced by separating the different wavelength components of white light with a prism. He used three similar mercury thermometers with bulbs blackened so that they would absorb radiation effectively. Two of these thermometers were used to monitor the ambient temperature and the third recorded the temperature when its bulb was placed in different regions of the spectrum. Herschel found that the temperature steadily increased as he went from blue to red, but he also discovered that the temperature continued to rise beyond the visible red end of the spectrum. He referred to this invisible part of the spectrum as *caloric radiation,* later to be called infrared radiation.

Modern thermal detection is based on devices that are more sensitive than a mercury thermometer. One such device, which basically measures temperature, is the *thermocouple,* based on what is known as the Seebeck effect, after its discoverer, the German physicist Thomas Johann Seebeck (1770–1831). If there is a thermal gradient in a conductor — a metal or a semiconductor — then a potential difference is established within it. However, the more usual form of thermocouple is illustrated in Fig. 6.3. In this device two

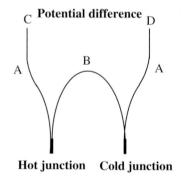

Figure 6.3   A schematic thermocouple for measuring temperature.

different conducting materials are used, designated as A and B in
the figure.

The hot junction is placed directly in contact with the temperature to be measured and the cold junction is kept at some standard
temperature. The measured potential difference developed between
C and D is a direct indication of the temperature difference between
the hot and cold junctions. The materials chosen for A and B, usually metals, depend on the range of temperatures to be measured.
For low temperatures, up to about 700 K, a common combination is
copper and constantan, the latter being an alloy of 55% copper and
45% nickel. While these give a good potential difference per unit temperature rise, to obtain an accurate measurement at higher temperatures it is necessary to use metals with a higher melting point. The
combination of tungsten with rhenium, with melting points 3,695 K
and 3,459 K respectively, can be used to measure temperatures up to
about 2,700 K.

An interesting related phenomenon to the thermoelectric effect
is the *Peltier effect* discovered by the French physicist Jean-Charles
Peltier (1785–1845). If a potential different is imposed on the system
in an opposite direction to that produced by the photoelectric effect
then heat is absorbed at the hot junction and given off at the cold
junction — essentially driving heat from a lower temperature region
to a higher temperature region. This is a refrigeration process that
has limited application in some scientific, and other, devices where
small-scale refrigeration is required.

The thermocouple is a device for measuring temperature but
what we are interested in is a device for measuring infrared radiation. If the hot junction of a thermocouple were exposed to a distant
heat source then, in principle, it would absorb radiation, heat up to
some extent and give a potential difference, indicating that infrared
radiation was present. In practice this would be a very insensitive
detector. However, the sensitivity can be greatly increased if several
thermocouples are joined together in series, as shown in Fig. 6.4, to
form a *thermopile*. Now the small potential differences generated by
each of the thermocouples add together to give a much larger output
potential difference. The incoming radiation falls on a heat absorber,

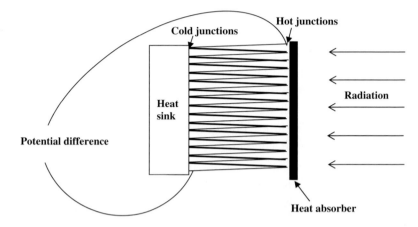

Figure 6.4   A schematic thermopile.

a blackened surface that heats up and gives the temperature of the hot junctions. The cold junctions are all connected to a heat sink, which takes in the heat conducted from the hot junctions and radiates it away. Balancing the rate at which heat is conducted and radiated maintains a difference of temperature between the hot and cold junctions that depends on the rate at which radiation falls on the heat absorber. The rate at which heat is conducted and radiated also affects what is known as the *time response* of the thermopile, i.e. the time from being exposed to radiation to when it gives a steady-state reading.

Antimony and bismuth give a particularly effective combination of conductors for thermopiles but modern commercial thermopiles are manufactured with the same kind of technology that is used for integrated circuits — photolithography and the deposition of thin films — with differently doped[3] silicon as the thermopile conductors. They can be very compact, with circular cross-sections a few millimetres in diameter. Their time response, the interval between being exposed and reaching a steady state, is usually in the range

---

[3]Doping a semiconductor means adding a small quantity of an element that changes its electrical properties. A full description can be found in Woolfson, M.M. (2010) *Materials, Matter and Particles: A Brief History*, London, Imperial College Press, pp. 245–50.

10–20 ms. The responsivity is normally between 1–$100\,\mathrm{VW}^{-1}$ with an NEP as low as $3 \times 10^{-10}\,\mathrm{W\,Hz}^{-\frac{1}{2}}$.

There are many commercial uses for thermopiles. When placed in a fixed position relative to the source of heat they can measure temperature, for example in a car, in a furnace or as an ear thermometer for body-temperature measurement. There are also space applications: a thermopile was one of the instruments in the Mars Climate Sounder, a component of NASA's Mars Reconnaissance Orbiter, launched in 2005. Thermopiles can also be used just as detectors of a source of infrared radiation and are installed as safety devices that detect either the absence of heat that should be present, as when the pilot light of a boiler is extinguished, or the presence of unwanted heat, as when a hairdryer overheats.

6.2.3. *Bolometers*

The bolometer, invented by an American astronomer, Samuel Pierpont Langley (1834–1906), is a device for measuring the rate at which electromagnetic energy falls on a surface. Even though its basic design is very simple, for some applications it is the most sensitive detector available. A simple bolometer is represented in Fig. 6.5. Radiation of total power Q (W) falling on an absorber maintains it at a temperature T (K). This absorber is thermally well insulated

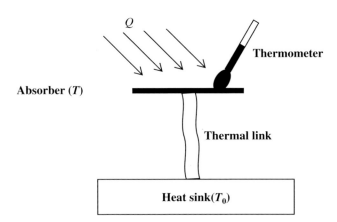

Figure 6.5   The basic structure of a bolometer.

except for a weak thermal link of thermal conductance G $(W\,K^{-1})$ with a heat sink, a region with a high heat capacity that can absorb a great deal of heat energy without an appreciable change of temperature, maintained at T_0 (K). The greater the rate at which energy falls on the absorber the greater its temperature and this temperature can be measured by means of a sensitive thermometer. The temperature difference between the absorber and the heat sink is given by

$$T - T_0 = Q/G. \tag{6.8}$$

Bolometers can measure the rate of arrival of electromagnetic energy over a wide range of wavelengths and are particularly valuable for astronomical investigations of very low temperature sources, such as star-forming clouds or disks around young stars, for which the peak emissions are within the wavelength range 100–1,000 μm. In this range bolometers are the most sensitive detectors of electromagnetic radiation but are bettered by other kinds of instrument for shorter wavelengths. However, a strength, but also a weakness, of bolometers is that they measure any incoming energy that can be absorbed by the instrument and apart from photons of electromagnetic radiation this can also include particles, both charged and uncharged. Thus, unless one can be sure from other kinds of measurement that particles are not a significant component of the incoming energy, it may not be known that just radiant energy is being measured. To give an idea of how sensitive a bolometer can be, in 1880 Langley had developed the instrument to the point that he could detect the heat radiated by a cow at a distance of 400 m; modern instruments are much better than that.

In Langley's original instrument the absorber was in the form of a blackened strip of platinum that formed one arm of a Wheatstone bridge, a device for measuring electrical resistance. The greater the temperature of the platinum the higher was its resistance so measurement of the resistance gave the temperature. In a modern bolometer the thermometry is often done by measuring the change of resistance of a strip of doped germanium that forms the absorber. The doped germanium is a semiconductor material for which, unlike for a metal, the resistance *decreases* with temperature. Semiconductors are materials that contain atoms with structural electrons that can be detached by the addition of thermal energy and can then move freely through the material as conduction electrons. The amount of energy required to detach an electron is expressed as kT_g, where k is the Boltmann constant, 1.38×10^{-23} J K^{-1} and T_g, expressed in K, is the *band-gap temperature*. The higher is the temperature of the semiconductor the greater is the number of conduction electrons produced and hence the *lower* the resistance. The relationship between resistance and temperature is approximately

$$R_T = R_0 \exp(-T/T_g). \tag{6.9}$$

For small T_g the change of resistance per unit change of temperature can be very large so providing a very sensitive and precise measurement of temperature. The actual arrangement, shown in Fig. 6.6, gives the change in resistance of the absorber in terms of a measured voltage. The load resistance, R_L, is much larger than the resistance of the absorber, R_T, so that, even although the resistance of the latter varies the current through it remains virtually constant. Thus the voltage across it, which is current × resistance, varies and measurement of the voltage gives the resistance and hence the temperature that, in its turn indicates the total power absorbed by the instrument. However, with the arrangement shown in the figure the current passing through the absorber is generating power, equal to the product of the voltage and the current. This power must be added to Q in (Eq. 6.8).

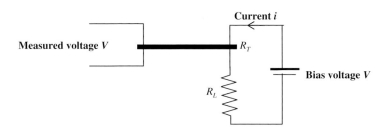

Figure 6.6   The temperature measuring arrangement for a bolometer.

For the highest bolometer sensitivity it is best that the changes of temperature due to the incoming radiation should be as high as possible relative to the ambient temperature and the ambient temperature should be kept as constant as possible. For very demanding astronomical measurements the instruments are cooled by liquid helium to a temperature in the region of $0.1\,\mathrm{K}$, which reduces random electronic (Johnson–Nyquist) noise in the circuit. At their best bolometers can give a responsivity of $100\,\mathrm{VW}^{-1}$ and an NEP better than $10^{-20}\,\mathrm{W\,Hz}^{-1/2}$ is attainable.

6.2.4. *Golay cells*

In its simplest form a Golay cell is a small cylinder containing gas, one metallic end wall of which is blackened to absorb radiant energy with the other end consisting of a diaphragm of plastic material. When the gas inside is heated it expands and distorts the plastic, the movement of which is detected by some optical means, such as the deflection or spread of a reflected beam of light. A schematic impression of a Golay cell is given in Fig. 6.7, although actual commercial devices are usually much more complicated, especially in their optics. In the arrangement shown, the more the flexible diaphragm is extended the greater is the spread of the beam reflected from its surface and the less the light that is falling on the photocell.

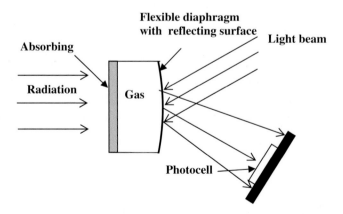

Figure 6.7   A schematic Golay cell.

Thus, the greater the temperature of the gas the less is the output from the photocell — a means of measuring the temperature. Operating at normal room temperature a well-designed Golay cell can give a responsivity of $1.5\,\mathrm{VW}^{-1}$ and an NEP of $10^{-10}\,\mathrm{W\,Hz}^{-1/2}$.

6.2.5. *Pyroelectric detectors; intruder alarms*

A pyroelectric material is one that is polarized, i.e. generates a temporary electric potential across itself by the motion of electric charge within it, when it is either heated or cooled. This charge quickly disappears during a period known as the *dielectric relaxation time*, which can be as short as nanoseconds (10^{-9} s) or less. If the temperature of the material remains constant, no matter at what level, then no polarization is developed. A pyroelectric material, such as lithium, with a blackened surface so as to absorb radiation effectively, would give no signal if just heated by some infrared source to a constant temperature. However, if the material is thin, with a low thermal capacity, then it will heat up and cool down quite rapidly with changes of incident radiation and this enables it to be a detector of infrared radiation by combining it with a *signal chopper*. This is a device, something like a cooling fan with separated blades, which intermittently blocks off the incoming signal, typically with a frequency in the range 25–60 Hz. Now the pyroelectric material is alternately heated and cooled and the change of polarization will enable the received infrared radiation to be both detected and to have its power measured, since the extent of the polarization depends on the rate of change of temperature. Another advantage of chopping the radiation is that it provides a modulated signal that is convenient for processing by AC circuitry to carry out processes such as *digital signal processing* to further improve the estimates of the source signal strength.

There are many advantages in the use of pyroelectric detectors. These are:

- Depending on the material, they can operate over a wide temperature range.

- They are sensitive over the complete range of the electromagnetic spectrum.
- They can respond rapidly, even in picoseconds (10^{-12} s).
- They require very low power to operate.
- Pyroelectric materials are normally of low cost.

Common devices that detect infrared radiation are the intruder alarm systems installed in many homes. They are activated by the infrared radiation generated by the heat of a human body but they must have rather special properties if they are not to generate large numbers of false alarms. For example, a particular room in which the alarm is installed may have a wide range of temperatures, perhaps becoming quite cool during a winter night when the central heating system is not operating to becoming quite hot either when the central heating switches on or on a hot summer day. The answer to this is to have a device that detects the *motion* of an infrared emitting body since that would distinguish an intruder from the stationary features of a room that vary in temperature.

A pyroelectric detector can be used to meet this requirement of distinguishing a stationary source of varying temperature (the electric fire) and a moving source of constant temperature (an intruder). The way of overcoming the problem is to have the detector consisting of a number, typically four, equal sections of pyroelectric material usually mounted on a chip that also contains all the electronic circuitry required by the system. There are small lenses each of which focuses a different part of the overall field of view on to one of the sections. These sections are so connected that their electric potential outputs cancel each other. They are all equally affected by variations of the general background radiation of the room so, no matter how this changes, the alarm will not be activated. However, radiation from an intruder entering a room is focused by one lens onto a section that was previously only irradiated by a cooler part of the room and the resultant section is polarized. As the intruder moves so the temperature change moves from one section to another and the electronics detects the resultant changes of polarization and triggers the alarm. In front of the pyroelectric sections there is normally a plastic material in which the lenses are moulded and which only transmit a range of wavelengths around 9.4 μm, the optimum wavelength for the emissions at the temperature of the human body, about 311 K.

6.3. Infrared Imaging

There are two important kinds of infrared imaging. The first can be generally categorized as *night vision*, the ability to see objects in the absence of visible light. Much of this development has been for military use in producing night-vision devices for either just seeing the environment or as a night-sight for rifles. Security forces, such as the police, also make extensive use of night-vision equipment in their operations against criminals and naturalists who wish to study nocturnal animals in their normal habitats normally use infrared cameras. The second kind of imaging is called *thermographic imaging* in which the purpose is to gauge the temperatures of various parts of the object being imaged.

Any imaging system requires two basic components — first, a means of creating an image, such as a lens, and second, a way of recording the image, either in an analogue form on film or in a digital form that can then be converted into a visual display. In some situations the object to be imaged may not itself be emitting enough infrared radiation to give a clear image, in which case it can be illuminated by an infrared source, say an infrared laser, and then seen by the radiation it reflects.

6.3.1. *A night-vision device*

In order to make effective lenses for a thermal camera it is necessary to use lens material that is transparent to a range of infrared wavelengths. A favoured material is synthetically made zinc selenide, ZnSe, that is transparent to infrared radiation over the very wide range of 0.5 to 20 μm. Another material often used is AMTIR1, a synthetic glass containing germanium, arsenic and selenium with chemical formula $Ge_{33}As_{12}Se_{55}$, with good transmission over the range 0.7 to 12 μm.

Night-vision devices operate with radiation that covers the range from visual wavelengths through to the near infrared. Clearly, under the conditions that such devices are used the visual part of the radiation is very weak and the majority of the radiation that forms the image originates as near infrared. Given that it is possible to produce an image for infrared radiation with a lens the problem is to produce from this weak low-powered invisible source image an image that is visible, intense and well defined. The way that this is done is by converting each photon that forms the source image, whatever its energy (wavelength), into many photons all with energy corresponding to the same visible wavelength. The essential part of this process is a device called an *image intensifier*, illustrated in schematic form in Fig. 6.8.

The first stage of the process is the formation of an image on the photocathode, a plate some 25 mm in diameter on which is deposited a thin layer of a mixture of alkali metals or a small band-gap semiconducting material. When a photon falls on the layer it causes the emission of an electron through the photoelectric effect. Some materials

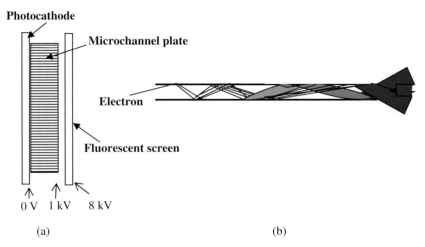

Figure 6.8   A schematic image intensifier. (a) The main components (b) The multiplication of an electron passing through a microchannel.

have very weakly attached electrons — for example, the valence electrons of alkali metals — and a sufficiently energetic photon, interacting with the material can give the electron enough energy to escape from the surface. For this reason at each point of the image on the photocathode there is a rate of release of electrons that is proportional to the rate at which photons are arriving. The *quantum efficiency* of the photocathode is defined as the proportion of incident photons giving rise to electron emission; for a commonly used photoelectric semiconductor, gallium arsenide (GaAs), this varies with the photon wavelength but is typically of order 40–50%. The electrons are propelled through the image intensifier, from left to right in the figure, by an applied electric field. Because of small amounts of gas contained in the intensifier, some positive ions are produced that the field sends from right to left so that they could impinge on the photocathode. The GaAs coating would deteriorate quickly under such bombardment so it is protected by a thin layer of a metallic oxide that is deposited on the inlet face of the microchannel plate, which prevents ions getting back to the photocathode but gives the penalty that it reduces the quantum efficiency by 30% or so.

The microchannel plate is the real high-technology aspect of the image intensifier. It consists of a slightly conducting glass plate, about 0.5 mm in thickness, which contains millions of densely packed very fine channels that have diameters as small as 6 μm. The insides of these channels are coated with a secondary electron emitter, such as caesium iodide (CsI), that gives off many lower-energy electrons if bombarded by a higher-energy electron. There is a potential difference of several hundred to one thousand volts across the microchannel plate so the electrons that enter are constantly accelerated and when one strikes the surface of a channel it gives rise to a number of lower-energy electrons. These in turn are accelerated and when they strike the wall each of them gives a number of lower-energy electrons. After several such collisions a single electron entering the plate gives rise to many thousands of emerging electrons (Fig. 6.8(b)).

The final stage of the intensifying process is that the electrons emerging from the microchannel plate are accelerated through

Figure 6.9   A night-sight image.

another few thousand volts and then strike a fluorescent screen, where the energy of the electrons is converted into photons, all corresponding to the same wavelength of light, often in the green region of the spectrum. In the case of a night-vision device the image produced on the fluorescent screen, optically inverted if required, can be viewed directly on the screen. However, if the intensifier is in a camera then the image can be projected onto a CCD (§5.6) and stored electronically.

This description of an image intensifier, which is at the heart of night-vision devices, is quite simplified and over the years, since the first work by German scientists in the Second World War, there have been huge technical advances. Figure 6.9 shows a night-sight image of two American soldiers taken during the Iraq war of 2003. Although this picture is of much lower quality than a photograph taken in normal light it meets military requirements.

Much higher quality images can be produced if the subject is illuminated by a source of infrared radiation. A comparison of what is seen with and without active illumination is illustrated in Fig. 6.10. The increase in quality with illumination is striking. However, although the unaided eye cannot see infrared illumination it can be detected by night goggles, as may be used by a military adversary, so it is not normally used in a military context.

Figure 6.10   A comparison of images with a night-vision device with and without active infrared illumination (www.ExtremeCCTV.com).

6.3.2. *Thermography: thermal imaging*

The purpose of an infrared camera, or night-sight viewer, is to produce a visual image resembling, as far as possible, one that could be obtained with visible light. By contrast, the purpose of *thermography* is to detect and record the temperature distribution in the field of view using infrared emission of wavelengths typically in the range 0.9 to 14 μm. A *thermograph* often gives a rather coarse image of the object being observed but forming a well-defined image is not the prime purpose of the process.

Thermography depends on the principle that the hotter an object is then the greater is its output of heat energy per unit area; that is a consequence of (Eq. 6.1). The type of thermal imager being described here is based on the bolometer, described in §6.2.3 with the detector in the form of a *microbolometer array* (MBA) — tiny bolometers, each a few tens of microns in diameter and each of which constitutes a pixel of the imaging system. These tiny bolometers with the associated circuitry for each microbolometer, the equivalent to what is shown in Fig. 6.6, are created on the MBA by the kind of etching processes that is used to produce microcircuit chips. Most MBAs are either 160×120 pixels or, of better quality, 320×240 pixels. Larger arrays (640×480) are also available but are very expensive.

The fabrication of a MBA is, like other modern microchip manufacture, a marvel of modern industrial production. The net result of this selective etching and removal of material is illustrated in Fig. 6.11, which shows two neighbouring microbolometers.

An image is projected onto the MBA and each semiconductor absorbing plate heats up by an amount that depends on the rate at which energy is absorbed. Since some of the radiation passes through the plate without being absorbed a reflector below the plate returns this radiation back to the plate to give further absorption. The connector shown both gives electrical connection to the circuitry on the

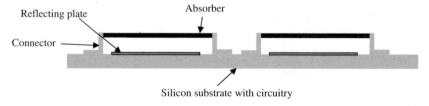

Figure 6.11   Two neighbouring microbolometers of a MBA.

silicon substrate so that the change of resistance of the absorber can be measured but also provides a thermal link to the substrate, which acts as a heat sink — as shown in Fig. 6.5. A typical rise of temperature of the absorber is about 0.03 K for every 1 K increase in the temperature of the region of the source being imaged at that microbolometer.

Each microbolometer output is recorded and then the data may be processed and output in any desired form. In one form of output the final image is in shades of grey, from white to black, with white indicating the highest temperature and black the coldest. This output can either be continuously varying or in steps of shades of grey. However, since the human eye is better at discriminating colour than shades of grey a more common form of a *thermogram* is one in colour. Figure 6.12 shows a *thermogram* of a dog, with a scale at the side so that colour may be interpreted as temperature. With suitable modifications to give a short response time it is possible to produce a thermographic camcorder that will record changes of temperature with time in the field of view — something needed for some applications.

Thermographic imaging has a wide variety of uses. Infrared radiation can penetrate smoke so enabling fire fighters to view the interior

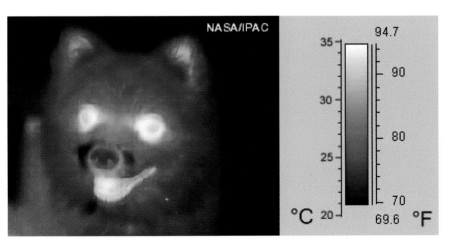

Figure 6.12   A thermogram of a dog with a colour-temperature conversion scale.

of a fire and detect the main locations of the highest temperature. It can also be used in various security situations — for example, it is carried by police helicopters that can locate and follow fleeing criminals by their heat emissions even on a dark night with little or no ambient light. If working machinery is viewed with a thermographic camera then the first signs of breakdown can be detected by spotting a high heat output, and hence rise of temperature, in a part experiencing excessive friction. An important use of thermography is in medicine as a diagnostic tool; many medical conditions can be detected by the development of either hot or cold regions in the body. In 2009, during an influenza pandemic alert, some airports installed thermographic equipment to detect passengers with a higher than expected temperature and who therefore might have been infected.

Chapter 7

Radar

7.1. The Origin of Radar

In 1865 the Scottish physicist James Clerk Maxwell made his greatest contribution to science, the theory of electromagnetic waves, and showed that light was such a wave motion. Then in 1887 the German physicist Heinrich Hertz (1857–1894) demonstrated the existence of radio waves, electromagnetic waves of longer wavelength than infrared radiation. It is difficult to define a division between the two kinds of radiation; any wavelength less than a millimetre can almost certainly be described as infrared while anything greater than a centimetre may certainly be considered as a radio wave.

If an emitted radio wave has its amplitude modulated by a sound wave, say that of music or a voice, then, with suitable detection equipment, the modulation can be heard as sound — this was first demonstrated in 1911 when a station in California started transmitting radio programmes. Others followed in the Netherlands, Canada and Argentina and the first national broadcaster, the British Broadcasting Company Ltd, now the British Broadcasting Corporation (BBC), was founded in 1922.

The fact that radio waves could be scattered and reflected by objects, particularly if they were metallic, was established quite early and, in 1904, a German inventor Christian Hülsmeyer (1881–1957) registered a patent for detecting the presence of metallic objects, e.g. a ship, by reflected radio waves. Over the next thirty years, in various countries, particularly in the UK, Germany, France, the USA and the Soviet Union, developments were made to use radio detection for military purposes.

The greatest advances in this field prior to the Second World War were made in Britain. There was a suspicion in Britain that German scientists were working on a death-ray device based on radio waves. The British research, led by Robert Watson-Watt (1892–1973) eventually reported that such a device was not possible but also concluded that radio waves could be used to detect and locate incoming aircraft; in the mid-to-late 1930s the prospect of war with Germany seemed very likely and the need to defend against possible air raids was pressing. The original British name of the project to develop an aircraft detection system was Range and Direction Finding (RDF), but the name used by the Americans, RAdio Detection And Ranging (RADAR) is now universally used and *radar* has become a standard English word.

Although the initial applications of radar were for purely military purposes it has developed into a versatile tool, many of the present applications being of a purely civil nature. These include:

Air traffic control	Sea traffic control
Speed traps	Intruder alarms
Navigation	Weather monitoring
Determining sea state	Locating archaeological relics
Distance and rotation of planets	Imaging planetary surfaces
Determining Earth resources	Mapping

The military applications have expanded from those originally envisaged and now include:

Detection of enemy or own forces	Weapon guidance
Detection of missiles	Espionage

We shall begin by considering the original purpose of radar systems, the detection of hostile aircraft or naval vessels. In this role, the basic requirement of a successful radar system is that it should be able to detect the direction and distance of an object, in essence its precise position, and a secondary aim is to determine its velocity, i.e. the direction and speed of its motion.

7.2. Determining the Distance

If a radio wave, reflected from a target, is detected at the same location as the transmitted signal then the time interval between the emission of the wave and the detection of the reflected wave is that required for the radiation, travelling at the speed of light, to travel twice the distance to the target (Fig. 7.1).

If the emitted wave were continuous, then the reflected wave would also be continuous and there would be no way of determining that time interval. The way to overcome this problem is to send out the radio wave as a succession of short pulses as illustrated in Fig. 7.2. This pattern of output is characterized by the duration of each pulse, τ, and the time interval between pulses, T, defined by the *Pulse Repetition Frequency* (PRF), the number of pulses emitted per second. In a typical case we might have $T = 1\,\mathrm{ms}$ ($10^{-3}\,\mathrm{s}$; PRF $= 10^3\,\mathrm{Hz}$) and $\tau = 1\,\mu\mathrm{s}$ ($10^{-6}\,\mathrm{s}$). The distance that electromagnetic radiation travels in time $10^{-3}\,\mathrm{s}$ at the speed of light, $3 \times 10^8\,\mathrm{ms^{-1}}$, is 300 km

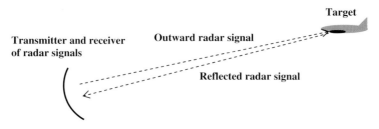

Figure 7.1   The distance travelled between the transmission and detection of a radar signal is twice the distance of the target.

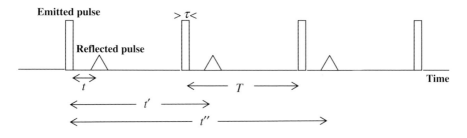

Figure 7.2   A pulsed train of radar signals with reflections.

corresponding to a radar range of 150 km. If the target is at a distance of 30 km then the return reflection will be 200 μs behind the outward pulse. Measuring this delay time, t, then gives the distance of the target.

If it were not known that the distance of the target was below 150 km then there is an uncertainty in the determination of distance. The delay in the return pulse could be t' or t'', corresponding to target distances of 180 km and 330 km respectively — and there are other possibilities. In most situations this ambiguity would not occur but if it does then it can be resolved by using a number of different PRFs. If a burst of pulses is sent out with a distance 200 km between successive pulses (PRF = 1,500 Hz) and t, as shown in Fig. 7.2, is 200 μs then the target distance could be 30 km or, in general $(30+100n)$ km with any positive integral value of n. If a following burst of pulses has a distance 360 km between successive pulses (PRF = 833.3 Hz) and t is 333.3 μs then the target distance could be $(50 + 180\,m)$ km with any positive integer as m. For $n = 2$ and $m = 1$ the distances are consistent at 230 km and this is indicated as the actual target distance. A third burst of pulses with yet another PRF could strengthen and confirm this estimate. An electronic system associated with the radar installation can easily solve the logical problem of finding the correct distance.

7.3. The Basic Requirements of a Radar System

An effective radar system requires a number of basic components. These are:

(i) The radar signal requires a source that gives high-powered radio frequency (RF) short pulses. The higher the power of each pulse the greater will be the reflected signal and the ease with which it can be detected. The shorter the pulse the better will be the determination of distance. A pulse one microsecond in duration has a length of 300 m and this will be the order of magnitude of the accuracy of distance determination using pulses of such

length. For some purposes, i.e. air traffic control, where precise distance estimates are critical for safety requirements, even
shorter pulses are required — although, as we shall see in §7.9.1,
alternative air traffic control systems have been developed to
achieve that purpose.

(ii) The generated pulses must be sent to an antenna. This requires
the use of a waveguide, usually in the form of a hollow metallic
tube of rectangular cross section.

(iii) An antenna that radiates the pulses in the form of a beam with
some required intensity distribution.

(iv) A means of locating the target by scanning the beam. There are
both mechanical and electronic methods of achieving this.

(v) A receiver to detect the reflected signal. This also requires the
use of an antenna, which can either be quite independent of
that which transmits the signal or the same one that is switched
between the two modes of operation.

(vi) Interpretation of the reflected signal and presentation to the
user of the information in a useful and readily understood form.

7.4. Generators of Radio Frequency Radiation

Any length of wire carrying alternating current will be a source of
electromagnetic radiation with a frequency equal to that of the current. However, a single vertical linear wire sends out equal power in
all horizontal directions and, with such a source, it would only be
possible to determine the distance of the target but not its direction,
and hence its actual position. Another feature would be that sending
the signal out equally in all directions is very wasteful, in the sense
that most of the energy would be achieving nothing and a reflected
signal would be very weak. If the energy were concentrated in a narrow beam that hits the target then the direction of the target would
be determined, the intensity of the beam would be greater and there
would be a much stronger reflected signal. This objective is achieved
by antenna design by which in a transmitting mode it converts an
alternating electric potential into electromagnetic radiation and in a
receiving mode it converts incoming electromagnetic radiation into

an alternating electric current. It is clear that for an effective radar detection system what is required is a powerful source of RF radiation and an aerial that will concentrate the energy into a narrow beam.

There are two main types of generators of RF electromagnetic radiation, the *klystron* and the *cavity magnetron*, and these are now described.

7.4.1. *The klystron amplifier*

A schematic representation of a klystron amplifier is shown in Fig. 7.3.

In the klystron a beam of electrons produced from a heated cathode is accelerated by the electric field between the cathode and anode. This beam is passed into a cavity in which low-power RF energy, tuned to the cavity's resonant frequency, is fed in, which causes the electrons to form bunches. The bunched electron beam then passes into a second cavity in which they generate RF *standing waves* with the same frequency as that input into the first chamber. In the output cavity the RF wave energy generated is channelled out via a *waveguide* to be picked up by an antenna and then reradiated into space in the required form. The overall effect is that the electrical energy used to accelerate the electrons efficiently amplifies the low power RF radiation fed into the bunching cavity into the high power RF radiation being extracted from the standing-wave cavity.

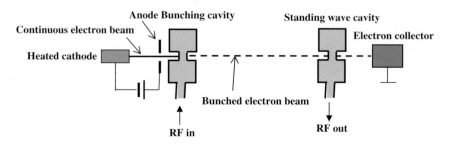

Figure 7.3   A two-cavity klystron amplifier.

7.4.2. *The cavity magnetron*

The cavity magnetron, developed by the British physicists J.T. Randall (1905–1984) and H.A. Boot (1917–1983) in 1940, made an important contribution to the effectiveness of radar during the Second World War. It works on a different principle from the klystron and is capable of developing high power with high frequency in a small device, although at the expense of poorer frequency control. A schematic representation of a cavity magnetron is shown in Fig. 7.4.

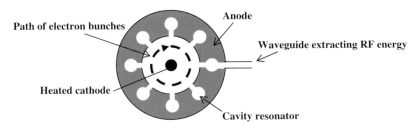

Figure 7.4   A schematic cavity magnetron.

The cathode is in the form of a conducting rod, coated with a material that emits electrons when heated. This cathode is subjected to a high negative potential from a direct current source, which can either be continuous or pulsed. The cathode is coaxial with the evacuated cylindrical anode, kept at ground potential, within which are situated the cylindrical resonant cavities. A permanent magnet creates a magnetic field parallel to the axis that, together with the radial electric field, causes the electrons to move in a circular path. As they sweep by the cavities they induce the production of high-frequency radiation fields within them; this phenomenon is similar to that of producing a sound within a bottle, at a frequency characteristic of its volume and shape, by blowing across its neck (try it and see). Just as for the bottle, the frequency of the radiation field depends on the dimensions of the cavities, and this reacts back on the stream of electrons causing them to form bunches. Some of the RF energy is extracted via a waveguide. For radar applications an antenna takes up this energy but the cavity magnetron is also the source of radiation in the domestic microwave cooker, in which case the radiation passes directly into the cooking chamber.

For the production of pulsed RF radiation of between 8 and 20 cm wavelength — the so-called S-band — peak power of up to 25 MW $(2.5 \times 10^7 \, \text{W})$ can be produced.

7.5. Transmitting the Pulses

Once high-powered RF radiation has been produced, passing through a waveguide, the next stage is to transmit it in the form of a narrow beam. This is the function of an antenna.

7.5.1. *A simple dipole*

A typical dipole aerial is illustrated in Fig. 7.5(a). It consists of a metal rod divided at the central point at which point the RF signal is fed in. For optimum operation the overall length of the aerial should be about one-half of the wavelength of the radiation being emitted or detected.

The emission from a dipole aerial is illustrated in Fig. 7.5(b). In the horizontal plane the emission, shown as a circle, indicates equal emission in all directions. In the vertical plane the strongest emission is in the plane perpendicular to the aerial with another smaller maximum of emission along the direction of the aerial.

7.5.2. *The parabolic reflector*

A simple way of concentrating the radiated signal is with a *parabolic dish* (Fig. 7.6).

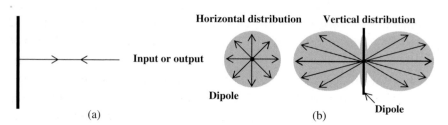

Figure 7.5   (a) A simple dipole aerial and (b) the emission from it in the horizontal and vertical planes.

Figure 7.6   A C-band (3.75–7.5 cm wavelength) parabolic radar dish (NASA).

A parabola is a curve defined by the equation $y^2 = x$ and is shown in Fig. 7.7. There is a point called the *focus* situated on the x-axis at $(1/4, 0)$ such that radiation leaving the focus in any direction and striking the parabola, regarded as a mirror surface, will be reflected parallel to the x-axis.

The shape of the parabolic dish shown in Fig. 7.6 is that of a surface generated by rotating the parabola shown in Fig. 7.7 about the x-axis. The bars coming outwards from the dish are supporting the antenna at the focus so that any radiation leaving the focus and travelling towards the dish is reflected in a tight beam along the dish axis. This gives a pencil-like radar beam that, once it homes in on the target, defines its direction with good precision. The larger is the dish the less is the angular spread in the beam it produces.

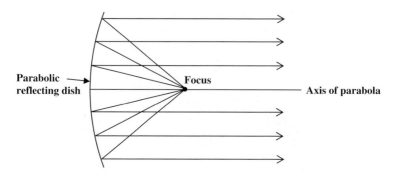

Figure 7.7   A parabola showing the reflection of rays coming from the focus.

The dish shown in Fig. 7.6 has a smooth continuous surface that has to be accurately profiled to a small fraction of a wavelength — corresponding to an accuracy of a few millimetres in this case. However, perfectly acceptable dishes can be constructed for which the surfaces are not continuous but have a cage-like structure, as long as the gaps in the structure are less than about one-tenth of a wavelength. Such dishes have the advantage of being lightweight and of offering lower resistance to high winds.

Radar dishes can be steered both in azimuth, i.e. in a horizontal plane, and in altitude to centre on the target. This is perfectly satisfactory in most civilian applications, such as air traffic control where the approximate direction of the target is predictable and the target wishes to be found and is cooperating with the radar system. In other cases, such as in the approach of hostile aircraft, the approach path may be completely unknown and the target will certainly not be cooperating in the attempt to detect it. In such cases ease of detection is more important than precise location and a beam that has a small spread in the horizontal direction and a much larger spread in the vertical direction makes the search process simpler and quicker because it just requires a horizontal scan. Such a fan-shaped beam can be produced with what is known as a *spoiled parabolic dish* (Fig. 7.8). This arrangement gives good azimuthal definition but little information about height. If the fan is oriented so that the beam width is narrow in the vertical direction and wide in a horizontal direction then, conversely, by moving the dish in what is known as

Figure 7.8   A spoiled parabolic dish with a network structure.

a nodding mode, i.e. up and down, the height of an aircraft can be determined but not its azimuthal position.

7.5.3. *Multiple-dipole-array antennae*

Although the parabolic dish is commonly used to produce a tight radar beam, this can also be achieved by the use of an aerial array. The simplest array is illustrated in Fig. 7.9 that shows two parallel dipoles, D_1 and D_2 a distance d apart. They emit equally with their emissions in phase, meaning that the peaks and troughs of emission happen at the same time in each of them.

In direction 1, perpendicular to the line joining the dipoles, at some distance from the array the emission is received in phase and the amplitude of the total signal is just twice that from each dipole. However, in direction 2, making an angle θ with direction 1, there is a path difference of

$$D_1P = d\sin\theta \qquad (7.1)$$

and if this distance is a whole number of wavelengths then the beams will reinforce each other in that direction. The condition for this to be true is

$$d\sin\theta = n\lambda \qquad (7.2)$$

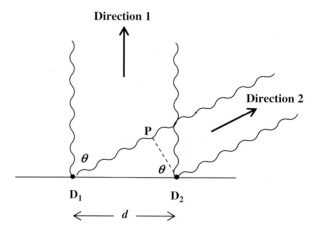

Figure 7.9   A two-dipole array.

where n is an integer. By contrast, if the path difference is such that

$$d \sin \theta = \left(n + \frac{1}{2} \right) \lambda \qquad (7.3)$$

then the peaks of one emission correspond to the troughs of the other and they cancel each other to give zero net emission.

A convenient visual way to represent these relationships between the two beams is by means of a vector diagram, known as an *Argand diagram*, as shown in Fig. 4.15. A vector, the length of which gives the amplitude of the beam and the direction its phase, represents the contributions of each beam. These vectors are connected end-to-end to give the amplitude and phase of the resultant. The cases represented by (Eq. 7.2) and (Eq. 7.3) are illustrated in Figs 7.10(a) and 7.10(b). When the emissions differ in phase by $60°$ the resultant is shown in Fig. 7.10(c).

The intensity of the beam is proportional to the square of the net amplitude and its variation with direction in the horizontal plane depends on the separation of the two dipoles. This is shown in Fig. 7.11 for $d = \lambda, \lambda/2$ and $\lambda/4$. For a separation of one wavelength the peak in the forward direction ($\theta = 0°$) is fairly narrow but there other peaks of equal intensity at $\theta = \pm 90°$. Making the

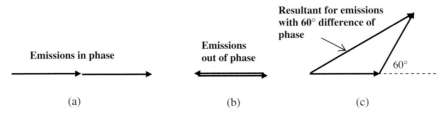

Figure 7.10   Resultant for (a) two in-phase emissions, (b) two out-of-phase emissions (slightly displaced for clarity) and (c) two emissions with 60° phase difference.

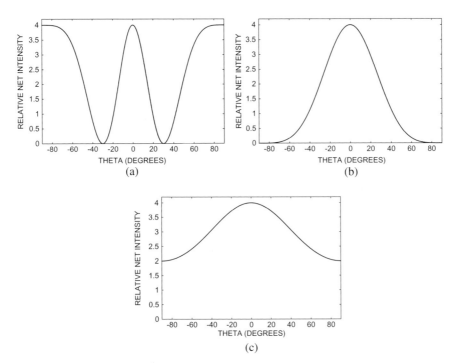

Figure 7.11   Net intensity distributions with angle for dipoles with separations; (a) one wavelength, (b) half wavelength and (c) quarter wavelength.

separation equal to a half wavelength gives a single peak in the forward direction but at the expense of having a much broader peak that would be less able to define the direction of the target. Having a lesser separation of $\lambda/4$ gives a single peak but one that is even less sharp.

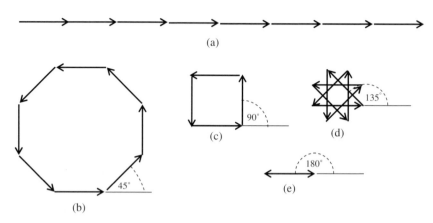

Figure 7.12   The resultant amplitude for a 8-element array with (a) $\phi = 0°$, giving the central maximum and zero resultants with (b) $\phi = 45°$, (c) $\phi = 90°$ where the vectors form two overlapping squares, (d) $\phi = 135°$ and (e) $\phi = 180°$ where the 8 vectors move alternately to the right and to the left to give a zero resultant.

The ideal situation would be to have a single sharp peak in the forward direction with no significant peaks elsewhere; this can be achieved by having a multi-dipole array with half a wavelength separation between neighbouring dipoles. First we consider the situation with an array of 8 dipoles by looking at the various Argand diagrams displayed in Fig. 7.12. In Fig. 7.12(a) we see the situation in the forward direction with the output from all the dipoles in phase. As θ increases so the phase difference between the output from each dipole and the next increases and the set of vectors curls up. When the difference, ϕ, reaches 45° the vectors have formed an octagon, shown in Fig. 7.12(b) and the resultant intensity is zero. As ϕ is increased, the next time a zero resultant is obtained is with $\phi = 90°$ (Fig. 7.12(c)) then with $\phi = 135°$ (Fig. 7.12(d)) and finally with $\phi = 180°$ (Fig. 7.12(e)). Between these minima of intensity there are obviously points of maximum intensity but the positions of these are most easily shown by computation. The intensity distribution for $d = \lambda/2$ and an 8-element dipole array is shown in Fig. 7.13(a) and for a 20-element array in Fig. 7.13(b). It will be seen that the larger array gives a narrower central peak and more subsidiary peaks, which are slightly smaller relative to the central peak.

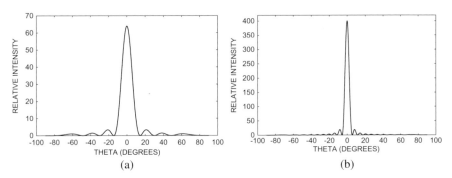

Figure 7.13   The resultant intensity for (a) an 8-element dipole array and (b) a 20-element dipole array.

7.5.4. *Phased-array radar*

The properties of the multiple-dipole-array antennae described in the previous section have been predicated on the dipoles emitting their individual signals in phase with each other. However, with an electronic circuit that enables controlled delays to be introduced to the emissions of individual dipoles it is possible to arrange matters so that signals from neighbouring dipoles have a constant phase difference, say ϕ. Such a system is called a *phased-array antenna*. Now the directions of the various maxima and minima will be different. In Fig. 7.9 we now consider the two neighbouring array elements, separated by a distance $d = \lambda/2$, emitting signals such that the phase of that from D_1 is ϕ ahead of that from D_2. Now the condition that the emissions at an angle θ to the normal should be in phase is

$$2\pi \frac{d \sin \theta}{\lambda} = \phi \tag{7.4}$$

and with $d = \lambda/2$ this gives

$$\theta = \sin^{-1}\left(\frac{\phi}{\pi}\right). \tag{7.5}$$

For $\phi = \pi/2$ this gives $\theta = \sin^{-1}(1/2) = 30°$.

From Fig. 7.13 we might expect that for a phased array with element separation $\lambda/2$ and with a phase difference of $\pi/2$ there would be a single large peak at 30° to the normal with smaller subsidiary peaks. This is indeed so and Fig. 7.14 shows the calculated emission at various angles for such a phased array of 32 dipoles.

The angle θ for the peak emission depends on the phase difference ϕ, the dependence for $d = \lambda/2$ being shown in Fig. 7.15.

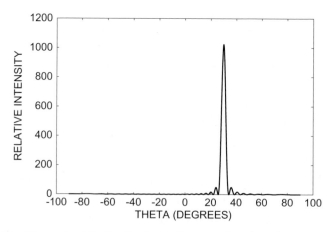

Figure 7.14   The output in the horizontal plane of a phased array with 32 elements separated by $\lambda/2$ and with a phase difference of $\pi/2$. in emission from neighbouring elements.

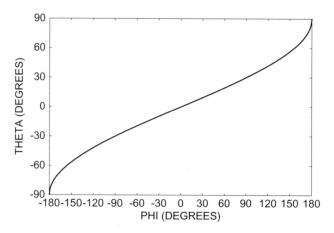

Figure 7.15   Variation in the direction of peak emission for a phased array with the phase difference in signals from neighbouring elements.

As ϕ varies from $-180°$ to $180°$ so θ sweeps out a complete arc from $-90°$ to $90°$. Thus it is possible by changing ϕ electronically to sweep the radar beam without any mechanical system being involved. This has the considerable advantage that the sweeping speed can be at a rate of thousands of degrees per second, which no mechanical system could match, and in each sweep of the beam several targets could be tracked.

Phased array radar is most commonly used for military systems for which the reliability of not having mechanical moving parts is a primary attraction. These systems often consist of a two dimensional array of dipoles in which electronically varying different phase differences between neighbouring elements along the x and y directions enable a fast scan in a raster mode to cover the whole potential target region.

Although it is possible to scan a range of angles from the normal from $90°$ to $-90°$ this is not done in practice. The reason for this is that the width of the peak becomes unacceptably wide at larger angles to the normal. Figure 7.16 shows the calculated widths of peaks for a range of angles for a 32-element array and it will be seen that close to $90°$ the peak is very broad. For this reason it is

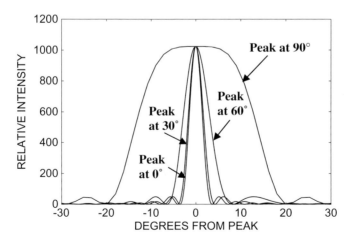

Figure 7.16   The peak width for different angles to the normal for a 32-element array.

Figure 7.17   A phased array radar installation in Alaska.

customary to restrict the range of scanning angles and to use several
arrays in a single installation to cover the whole of space.

Figure 7.17 shows an installation in Alaska designed to detect
intercontinental ballistic missiles. There are three arrays in the installation and the normal to each array is inclined at 30° to the horizontal. With this system scans of up to 60° from the normal in both the
horizontal and vertical directions can cover the whole of space. For
such a large array very precise targeting of small objects is possible;
not only can missiles be accurately tracked but many of them can be
detected simultaneously — a vital characteristic to defend against
multi-missile attacks.

7.6. Reception and Presentation

An antenna that is designed to create a signal from an alternating
current passing into it can also be used to receive electromagnetic
radiation and convert it into an alternating current. It is quite common to use the transmitter as a receiver by electronically alternating
the mode of use. There are technical problems that arise because
the output signal is usually at megawatt intensity and the reflected

Figure 7.18   A US Air Force Stealth Bomber.

signal at small fractions of a watt. However, electronic solutions to this problem are available. The time interval between the emitted pulse and that reflected gives the distance of the reflecting object and the intensity of the reflected beam a measure of how well the object reflects the radiation — which depends on its size, shape and the reflectivity of its material. For example, the shape and surface material of the Northrop Grumman B-2 Sprint aircraft, generally known as the 'Stealth Bomber' (Fig. 7.18), are designed so that it reflects radar waves very poorly and so it is difficult to detect.

A common form of presentation is on a *plan position indicator*, or PPI, illustrated in Fig. 7.19, that shows a radar picture of a storm system where the reflecting objects are raindrops in the clouds. On the screen display the radar antenna is at the centre and circles are drawn at intervals of 40 km. The colour coding indicates the concentration of rain within the storm cloud; the concentration increasing as the colour goes from green to orange. A single image cannot give a three-dimensional view of the storm system but this can be obtained by taking several such images, each looking at a plane view at different elevations from the horizontal.

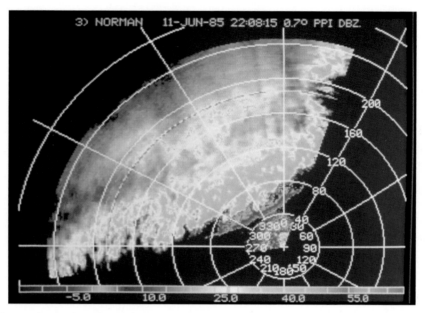

Figure 7.19 A radar image of a storm system (National Oceanic and Atmospheric Association).

7.7. Doppler Radar

We have seen how, by sweeping a radar beam either mechanically or electronically and measuring the delay in receiving the returning signal, it is possible to determine the position of the target. However, very often it is also necessary to determine the velocity of the target, i.e. its speed and direction of motion. An obvious way to do this is just to find its positional change over a known time interval but to get a good estimate the time interval might have to be fairly long — which may have security or safety implications. Clearly what is desirable is some way of determining the velocity at the same time that the position of the target is being found.

7.7.1. *The Doppler effect*

The phenomenon, known as the Doppler effect, was first described by the Austrian physicist Christian Doppler (1803–1853) and is a matter of everyday experience. The siren of an emergency vehicle is heard at a higher pitch (frequency) when the vehicle approaches and at a lower pitch when it recedes, where the higher and lower are relative to the pitch when the vehicle is at rest with respect to the listener.

As long as the speed of the vehicle is just a small fraction of the speed of sound in air, which is $1235\,\mathrm{km\,h^{-1}}$, then according to Doppler's equation the change of frequency df is related to the rest frequency, f, and the speed of the vehicle, v, by

$$\frac{df}{f} = -\frac{v}{v_s}, \tag{7.6}$$

in which v_s is the speed of sound. The sign convention for speed is that it is positive if motion of the emitting object is away from the observer and negative if towards. Thus df is negative for a receding sound source, i.e. the pitch is lower.

The wavelength of a sound wave, λ, is related to the frequency by

$$\lambda f = v_s \tag{7.7a}$$

or

$$\lambda = \frac{v_s}{f}$$

that on differentiating give

$$\frac{d\lambda}{df} = -\frac{v_s}{f^2} = -\frac{\lambda}{f} \tag{7.7b}$$

so that, again with the condition $v \ll v_s$, combining (Eq. 7.6) and (Eq. 7.7(b)) the change of wavelength of the received sound, $d\lambda$ is given by

$$\frac{d\lambda}{\lambda} = \frac{v}{v_s}. \tag{7.8}$$

The Doppler equation applies to the observations of wave motions of any kind, including electromagnetic waves. Thus the observed wavelengths of the spectral lines in the light emitted by receding stars are seen as of longer wavelength than if seen in a laboratory. Such light is said to be *red-shifted* since the observed wavelength moves towards the red end of the spectrum; similarly light seen from an approaching object is *blue-shifted.*

The above equations apply when the object being seen is itself the source of the wave motion. We are going to consider the radar situation where the observer is the source of the radiation and the object reflects that radiation back to the observer. In this case, in place of (Eq. 7.8) we have

$$\frac{d\lambda}{\lambda} = 2\frac{v}{c}, \tag{7.9}$$

where now we have now expressed the speed of light by the conventional symbol, c.

It must be emphasized that the only object speed that is involved in the Doppler equation is the component of the object's velocity along the line of sight (radial velocity). Transverse motion, at right angles to the line of sight, gives no Doppler effect and must be determined by some other means.

7.7.2. *Pulsed-Doppler radar*

Since the speeds of objects of interest to radar systems — even those of ballistic missiles — are tiny compared with the speed of light, $300,000 \, \text{km s}^{-1}$ or $1.08 \times 10^9 \, \text{km h}^{-1}$, Doppler-shift effects are very small and hence intrinsically difficult to measure. Radar operating at a frequency $3 \, \text{GHz}$ ($3 \times 10^9 \, \text{Hz}$) with short pulses of length $0.25 \, \mu s$ ($2.5 \times 10^{-7} \, \text{s}$) has 750 wavelengths in each pulse, which is of length $75 \, \text{m}$. If the target is approaching at $5,000 \, \text{km h}^{-1}$ then, according to (Eq. 7.9), the wavelength will be shorter and the length of the 750 wavelengths constituting a returning pulse will be different from that of an emitted pulse by $0.7 \, \text{mm}$ or 0.007 of a wavelength. Such a difference between an emitted pulse and a returning pulse would be difficult to discern; in fact the wave trains constituting the two pulses could overlap almost precisely and, if added, would for all practical purposes reinforce each other over the whole length of the pulse.

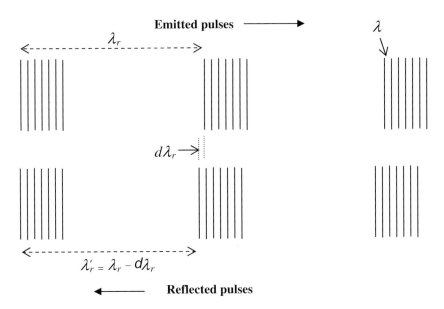

Figure 7.20   The streams of emitted and reflected pulses.

In a stream of pulses of emitted radiation there are two component wavelengths and frequencies involved. That of the radiation itself, say 3 GHz, we represent by λ and f. The other wavelength and frequency is that represented by the intervals, in space and time, of the succession of pulses — the PRF (§7.2), that we represent by f_r with corresponding wavelength λ_r. In Fig. 7.20 we show in schematic form a stream of emitted and reflected radar pulses. Both the radar wavelength and the *pulse repetition wavelength* are subject to a Doppler shift and, as seen in the figure, the latter is shortened by an amount $d\lambda_r$. As a *fractional* change of wavelength it is the same as that for the radar waves but since $\lambda_r \gg \lambda$ the absolute change is much larger. If the beginnings of the trains of emitted and reflected pulses are made to coincide, as in Fig. 7.20, then the displacement between the second pulses of the two trains is $d\lambda_r$ and if the two pulses added together they would do so with a phase difference of

$$\phi = 2\pi \frac{d\lambda_r}{\lambda} \tag{7.10}$$

If $\phi = \pm\pi$ and the individual pulses were of equal amplitude then the pulses would cancel each other to give a zero resultant; any other resultant would indicate the phase difference that gave it. In practice the phase difference is not found in this way. When the stream of pulses is emitted a record of it is kept and then the returning pulse train is also recorded. This information is manipulated mathematically to give the phase shift ϕ. Let us now see how the radial speed of the target, v, could be deduced from the phase shift. We have from (Eq. 7.9)

$$\frac{d\lambda_r}{\lambda_r} = \frac{2v}{c}. \tag{7.11}$$

Combining (Eq. 7.10) and (Eq. 7.11) and then using (Eq. 7.7)

$$\phi = \frac{4\pi v \lambda_r}{c\lambda} = \frac{4\pi v f}{cf_r} \tag{7.12a}$$

or

$$v = \frac{\phi c f_r}{4\pi f}. \tag{7.12b}$$

Now we carry out a typical calculation of speed using this equation. We take a radar system operating at a frequency of 500 Mhz (5×10^8 Hz) with a PRF of 10^5 Hz and a measured phase shift of $60° = \pi/3$ radians. This gives

$$v = \frac{(\pi/3) \times 3 \times 10^8 \times 10^5}{4\pi \times 5 \times 10^8} = 1{,}000\,\mathrm{ms}^{-1} = 3{,}600\,\mathrm{km\,hr}^{-1}. \tag{7.13}$$

A problem that arises with pulsed-Doppler radar is that if the value of ϕ strays out of the range $-180°$ to $180°$ then an ambiguity of speed arises since a phase shift of $60°$ is indistinguishable from one of $60° + 360° = 420°$ or many other angles. To keep ϕ within the desired range for large values of v it is necessary for the ratio f/f_r to not be too large. From the considerations of defining distance unambiguously, as seen in Fig. 7.2 it is desirable to have the distance between pulses to be as large as possible, which is to say having f_r small. Thus it appears that the conditions for measuring distance and speed unambiguously are somewhat incompatible but many radar systems can operate with variable f_r, depending on whether distance or speed is the prime consideration.

There are other ways of handling the radial speed ambiguity problem. Any given value of f_r will give a number of possible speeds, only one of which is correct. By taking measurements with two or more values of f_r it is possible to find the speed that they have in common, which will be the correct one. The rationale for this process is similar to that described in §7.2 by which the distance could be found unambiguously.

It must be stressed again that the radial speed found by pulsed-Doppler radar is just the component of the velocity of the target relative to the transmitter. The velocity, i.e. both speed and direction, can be found if radial speed estimates are made of the same target by two or more well-spaced transmitters. Alternatively, the transverse speed can be estimated by combining the range with the rate of change of azimuth angle, although this is an average speed over a small interval of time and not an estimate *at* a particular time.

7.8. Synthetic Aperture Radar

In our discussions of radar thus far, it has been used as a tool for locating a target and determining its velocity. Now we shall consider how it may be used for imaging.

We have previously commented on the fact that the limit of image resolution achievable with any radiation is of order of the wavelength so that with radar, at a frequency of 3 GHz the theoretical limit for resolution is about 10 cm. Equation 4.5 shows that for a microscope there is another controlling factor in resolution, the Numerical Aperture (NA), which governs how much of the radiation scattered by the object being imaged enters the imaging system. For a radar system, even one that is land-based and can use large receiving dishes, the effective NA is small and for an airborne system, where the size and weight of equipment is limited, it is extremely small.

The concept of synthetic aperture radar (SAR) is as follows. While it is impossible to have huge receiving equipment to combine the scattering over a wide angle, what can be done is to combine the signals from a number of well-spaced receivers. The image formed in this way will be of much higher resolution that that obtained from a single small receiver, albeit it will have a certain amount of noise since the radiation received is just a part of that scattered. The implication here is that the information being combined is that received *simultaneously* by the receivers. Carrying this idea one stage forward it is possible to have a single receiver that is moving and then to combine the signals received at *different times*. Such a situation could occur with an aircraft over-flying a region, emitting radar pulses and receiving the scattered radiation continuously as it does so.

In the most refined applications of SAR both the amplitude and phases of the scattered radiation are recorded continuously and the total information is then combined to give a detailed picture of the region flown over. However, to illustrate the principle that information taken at different times can be combined, a much simpler case will be described.

7.8.1. *A simple illustration of SAR*

In this application we consider an aircraft flying over flat terrain with sideways looking radar — radar that scans a wide area but only on one side of the aircraft's path. The plane emits radar pulses that illuminate the ground and the reflected radiation it receives is spread out over time with an intensity variation that depends on the reflectivity of ground features at different distances. If when the aircraft is at P_1 the intensity is strong corresponding to highly reflecting objects at distance R_1 then these objects could be anywhere on an arc at that distance from the plane, as illustrated in Fig. 7.21. At some later time, when the aircraft is at P_2, one of these objects will be at a different distance R_2 and again an arc defines the possible locations of the object. If this is repeated for a large number of returning signals for different positions of the aircraft then the arcs corresponding to this strong signal all overlap at the same point on the ground and locate the reflecting object.

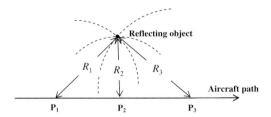

Figure 7.21 The arcs corresponding to strongly reflecting objects all intersect at one point.

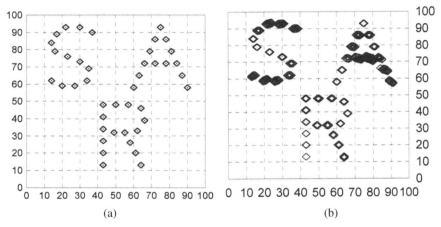

Figure 7.22 (a) A set of simulated radar radiation reflectors. (b) A simulated SAR image.

A computational illustration is now given of how this principle can give an image. In Fig. 7.22(a) reflecting objects, all with equal reflectivity, are placed on a grid. A table is set up corresponding to each grid point, initially with all table entries made equal to zero; in the computation both the x and y grid values were at unit intervals from 0 to 100 although a coarser grid is shown in the figure for clarity. Now an aircraft is imagined flying along the x-axis from the point $(0, 0)$ to $(100, 0)$ and at each grid point it emits a radar pulse. The returning signal comes from all reflecting objects; each reflecting object returns a pulse that indicates its distance but not its direction so the reflected information at each grid point is a net intensity at each distance from the aircraft. For simplicity all distances are expressed to the nearest grid-interval unit. If there is intensity I at distance 30 units then that means that there are I reflectors distributed on an arc at that distance from the aircraft. The aircraft position corresponds to a position in the table and each table point corresponding to a distance of 30 units from that position is then incremented by I. The net result of this exercise is shown in Fig. 7.22(b) as a contour map with the lowest contour at a height that eliminates most of the noise.

A real application of this principle would give a higher quality image than this rather crude illustration would suggest since the number of positions of the aircraft from which signals were sent and received would be much greater, so improving the statistical reliability of the results. The method could be extended to three dimensions in a non-flat terrain and it would also be able to record variations of reflectivity. It should be noted that the distances recorded are the slant distances to the reflecting objects, which depends on the height of the aircraft and allowance for this must be made in mapping ground features.

7.8.2. *More complex SAR applications*

In the simple illustration of SAR given in the previous section the only information that was used was the variation with time of the intensity of the reflected signal as the aircraft moved. However, it is possible to collect other information, for example the phase of the reflected signal or its state of polarization. The variations of the amplitudes of the electric and magnetic fields that characterize electromagnetic waves are perpendicular to the direction of motion of the radiation — they are *transverse waves*. When some objects reflect radar waves they may give different reflectivity for different polarization directions and this can be detected in the received signal. With very sophisticated radar transmitters and complex processing of the reflected signals, information can be extracted to give very detailed images. An example is shown in Fig. 7.23, which shows an SAR image of Venus, taken by the Magellan spacecraft through the planet's thick visually opaque atmosphere.

Figure 7.23   An SAR picture of the surface of Venus (NASA).

7.9. Other Radar Applications

In §7.1 the list of applications of radar shows what a useful tool it is in many aspects of civil, scientific and military life. Here two of these are briefly described.

7.9.1. *Secondary radar*

In the Second World War the use of radar to detect the approach of hostile aircraft and to detect and locate naval vessels, particularly submarines, was an important element in both the war in the air and in the Battle of the Atlantic. However, in the confusion of an ongoing battle the exact location of all the combatant units would not be known and one thing that radar could not establish was whether the target was friendly or hostile. Even in recent comparatively small-scale hostilities there have been incidents of deaths due to what is termed 'friendly fire' in which one set of military personnel attack some of their own side due to faulty identification.

Both sides in the conflict tackled this problem during the Second World War by the use of a system known as IFF (Identification Friend or Foe). Aircraft carried receivers that detected that they were under radar surveillance and responded by sending an encrypted radar signal that would positively identify them as friendly. Lack of response did not necessarily indicate hostility — the response equipment could be faulty or damaged — but no response triggered preparedness for a potential hostile attack.

The safe operation of a modern large and busy airport depends on the ability of air traffic control to know the exact location, including altitude, of all aircraft in the vicinity and to identify that aircraft. This is achieved automatically by a system known as SSR (Secondary Surveillance Radar). All aircraft, both commercial and military since they may be operating in the same airspace, are equipped with transponders, equipment that receives a radar signal on one frequency and transmits a response on a different frequency, which the radar transmitter is tuned in to receive. The response contains all the information required by air traffic control, identifying the aircraft, which a reflected radar signal could not do, and giving location and altitude with much greater precision than the primary radar systems being used could achieve.

7.9.2. *Ground penetrating radar*

It is a matter of everyday experience that RF electromagnetic radiation can penetrate visually opaque material. If this were not so then we should be unable to receive radio or television signals with indoor aerials. However, we also know that the depth to which penetration can take place is limited. A car radio loses the signal a short while after entering a tunnel passing through a hill or mountain. The degree to which the signal can penetrate depends on both the frequency of the radiation and the nature of the material. The absorption of the signal is due to the generation of currents within the material through which it passes and these are larger for high-conductivity materials. The lowest absorption, and hence the greatest penetration, is obtained with dry, porous material, for which the electrical conductivity is low and for longer wavelengths (lower frequencies). As examples, with the frequencies of radiation commonly used for Ground Penetrating Radar (GPR), around 200 MHz, the useful depth of penetration is a few centimetres for seawater, 10 m for clay, 100 m for limestone and 1 km for ice.

It is a characteristic of electromagnetic radiation that when it meets a boundary, on the two sides of which its transmission characteristics are different, then, even though the material on both sides is transparent to the radiation, some of it is reflected at the boundary. This happens with a glass lens when light falls on it due to the difference of the refractive index of air and glass, giving a difference in the speed of light in the two materials (§5.5). For radio waves the relevant property that varies from one substance to another and affects the speed of transmission of the waves is its *dielectric constant*; when a beam of radio-frequency radiation meets an interface between materials with different dielectric constants then some is reflected

Figure 7.24   A GPR survey being carried out in Wadi Ramm, Jordan.

backwards. The time between transmission of the radar pulse and detection of the reflected signal indicates the depth of the interface.

There are many different areas of application of GPR. In geological surveys it can delineate regions of subsurface rock and soil and disclose potential problem locations if, for example, a tunnel is to be excavated. In archaeology GPR can disclose the presence of the remnants of ancient constructions deep underground or the presence of burial chambers (Fig. 7.24). Although the name GPR may not seem appropriate here, it can also be used to explore the interiors of large extant structures like Egyptian pyramids or ziggurats built in the Middle East and South America. An everyday application is the use of GPR to find the location of pipes, sewers and other underground utility constructions when the exact location, or even whether they actually exist, is uncertain.

By moving the GPR equipment in raster fashion over an area and combining all the data found thereby it is possible to construct a three-dimensional image of what is below ground. The possible depth of exploration is greater at lower frequencies, but this has the penalty of giving lower resolution, so often the frequency chosen is a compromise between these conflicting requirements.

Chapter 8

Imaging the Universe with Visible and Near-Visible Radiation

8.1. Optical Telescopes

The first practical telescopes, which could magnify the appearance of distant objects by a factor of three, were certainly the work of Dutch lens-makers and opticians, although to whom the credit as inventor is due is not clear. Strong contenders for this honour are the spectacle makers Hans Lippershey, also credited with inventing the compound microscope (§4.3), and Zacharias Janssen (1580–1638), but the optician Jacob Metius (1571–1630) seems to have been developing the same idea at the same time. The date of the invention is usually quoted as 1608 and barely one year later, in July 1609, the English astronomer, Thomas Harriot (1560–1621) was viewing the Moon through a telescope and making drawings of what he saw.

The most famous of the early users of telescopes was Galileo Galilei (1564–1642; Fig. 8.1) who also observed the Moon in late 1609 and who developed much-improved telescopes, with magnifications of up to 30. From that time, telescopes were used for both astronomical and terrestrial observations; Galileo had a profitable sideline selling telescopes to merchants who could reap financial benefit by spotting returning merchant vessels long before they reached port.

Driven by mankind's strong desire to understand the nature of the Universe, over the centuries that have followed the design and construction of optical telescopes, especially for astronomical observations, have made great advances, the nature of which will be the topic of this chapter.

147

Figure 8.1   Galileo Galilei.

8.2. Refracting Telescopes

There is a story that Hans Lippershey was led to his discovery of
the telescope because children playing with lenses in his workshop
accidentally discovered that a large image of an object could be seen
if it was viewed through two lenses. The basic construction of a simple
refracting telescope — one with lenses as the sole components —
involves two lenses. The one closer to the object being viewed is
called the *objective* and the one through which the observer sees the
image is called the *eyepiece*. A simple refracting telescope, and the
way it produces an image, is illustrated in Fig. 8.2.

In astronomical use, and in most cases of terrestrial use, the
object is essentially at an infinite distance and hence the first inverted
image, as seen in the figure, is at distance f_o, the focal length of
the objective, from that lens. This image is at a distance from the
eyepiece equal to its focal length, f_e, at which point it produces a
virtual inverted image at infinity where it can comfortably be seen.

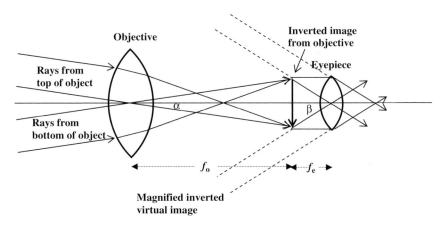

Figure 8.2   Image formation with a refracting astronomical telescope.

The object subtends an angle 2α at the viewing position and the image an angle 2β. For astronomical objects these angles are very small — even a large astronomical object like the Moon subtends an angle of only $0.5°$ — and the magnification produced by the telescope is given by

$$M = \frac{\beta}{\alpha} = \frac{f_o}{f_e}. \tag{8.1}$$

For most astronomical purposes the inversion of the image is not important. For terrestrial use, an extra lens is introduced between the objective and eyepiece that has the function of inverting the first image, restoring it to the correct orientation, without any change in its size.

Telescopes such as this gave a great fillip to astronomical observations in the seventeenth century but they did have severe disadvantages. The brightness of the image depended on the amount of light entering the telescope that, in its turn, depended on the size of the objective lens. Large lenses were difficult to make and were also quite heavy. However, of more importance were intrinsic problems associated with lenses. One of these is chromatic aberration (§4.3) requiring the use of composite lenses that are even heavier than simple ones. A second problem associated with lenses is that of

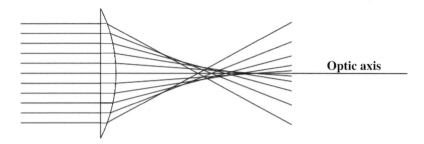

Figure 8.3   The effect of spherical aberration.

spherical aberration, due to the fact that rays at different distances from the optic axis come to a focus in different positions (Fig. 8.3).

The effect of spherical aberration can be minimized by suitably shaping the lens surfaces. This type of correction can also be made for *coma*, an aberration in which parallel rays not parallel to the optic axis come together to produce an image with a comet-like shape — a circular region with a tail.

8.3. Reflecting Telescopes

Imaging can be carried out not only with lenses but also with mirrors of convex or concave shape. The action of a concave mirror to produce a real image is shown in Fig. 8.4. When a light ray strikes the mirror the reflected ray leaves the mirror making the same angle with the normal to the mirror as the incident ray. This condition is independent of the wavelength of the radiation so there is no chromatic aberration. If the shape of the mirror is part of a spherical surface then there is spherical aberration just as for a lens. This can be removed by having a parabolic mirror (Fig. 7.7). The final aberration, coma, can be reduced by using a subsidiary lens system.

The advantage of mirrors over lenses in giving aberration-free optical systems was appreciated quite soon after the telescope was invented but it was not until 1688 that Isaac Newton (Fig. 3.1) constructed the first telescope using mirrors, of a type now called a *Newtonian reflector*. The telescope he constructed is illustrated in Fig. 8.5; the small plane mirror prevents some of the incoming light

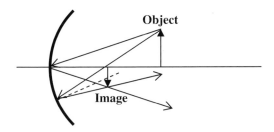

Figure 8.4   An image produced by a concave lens.

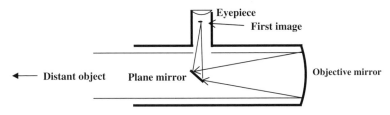

Figure 8.5   A schematic representation of Newton's 1688 telescope.

from reaching the objective mirror but this makes little difference to the quality of the final image.

The mirror material used by Newton was *speculum metal*, an alloy of two parts of copper to one part of tin. While it has fairly good reflectivity and gives an effective mirror surface it has the disadvantage of tarnishing so that it occasionally needs cleaning. The mirror had a diameter of about $1\frac{1}{3}$ in (34 mm) and a focal length of 6 in (152 mm). Although Newton understood the advantage of having a parabolic mirror the one he produced was spherical and hence much easier to grind. Light from the distant object was reflected back onto a plane mirror that deflected the light through an aperture in the barrel of the instrument to form an image. This was viewed with a lens eyepiece of very short focal length, about 5 mm. Newton's instrument gave a magnification of 30 or so and with it he observed the large Galilean satellites of Jupiter and the phases of Venus, both first observed by Galileo with his telescope. The telescope was about 15 cm in total length; Newton noted that it gave better magnification than an all-lens instrument some eight times longer. An adverse factor was that the reflectivity of speculum metal was much lower than that of a modern mirror so the image was darker than that produced by a refracting telescope of the same diameter.

Although it was not important for an instrument the size of Newton's, when it came to scaling up the size of telescopes, mirrors have many advantages over lenses. Not only can a mirror be much lighter than a lens of the same aperture but it can be supported all over its back surface, whereas a lens is normally just supported around its edge. The primary mirror of a large telescope is cast in glass, or some vitro-ceramic material, and then shaped by a grinding process. Finally the surface is aluminized, i.e. covered with a fine coating of aluminium about 100 nm thick. This process is accomplished by depositing aluminium vapour onto the glass surface in a vacuum enclosure. Since aluminium surfaces suffer from a slow process of corrosion they must be recoated at intervals of about 18 months.

There are many large telescopes in locations all over the world. The European Southern Observatory (ESO; Fig. 8.6) is located in Chile, taking advantage of the dry atmosphere of the Atacama desert. There are four instruments with primary mirrors 8.2 m in diameter. The light input from two or more of these, plus input from up to four

Figure 8.6   Part of the very large telescope complex in Chile.

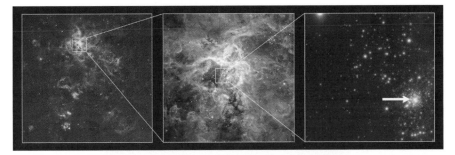

Figure 8.7   The position of the massive star R136a1 is indicated by the arrow in the final frame.

smaller 1.8 m instruments, can be combined, preserving the relative phases of the light received, so they form an interferometer that gives a resolving power 25 times greater than that of any individual instrument; with a resolution of 0.001" they could resolve the headlights of a car on the Moon! On 21 July 2010 a British team from Sheffield University announced that, using this facility, the Very Large Telescope (VLT) consisting of the four large telescopes, they had observed the most massive star yet discovered. This star, R136a1 (Fig. 8.7) is in a dense cluster of massive stars situated in a nearby galaxy, the Large Magellanic Cloud, at a distance of 165,000 light years. R136a1 is more than 250 times as massive as the Sun. Such a star has a surface temperature of 40,000 K, seven times that of the Sun, luminosity (light output) a million times the solar value and a lifetime of less than one million years, which makes the likelihood of observing such stars very low.

There are other individual large telescopes and assemblages of large telescopes in many locations — for example, in the Canary Islands, Hawaii, Australia and South Africa — the largest of them being more than 10 m in diameter. Figure 8.8 shows the 8.2 m Subaru instrument, situated in Hawaii and operated by the National Astronomical Observatory of Japan.

A limiting factor with imaging by a traditional ground-based telescope is refraction by the turbulent atmosphere. A star is essentially a point object and when turbulent cells of air, at different densities and temperature and hence with different refractive indices, deflect

Figure 8.8   The Subaru telescope showing the 8.2 m primary mirror.

light the point image moves in the field of view and gives the characteristic stellar-twinkle effect. Although planets often seem to be point-like they are actually small disks — of angular diameter 40" in the case of Jupiter — and despite small movements of all or part of the image a substantial part of the disk image remains always illuminated so the twinkling effect is greatly reduced. An observation that can be made with the largest traditional telescopes is a resolved image of the minor planet Pluto, at a distance of 6×10^9 km, and its largest satellite Charon, which is in an orbit of radius 19,640 km. Although the maximum angular separation as seen from Earth is just 0.7" the two bodies can be resolved (Fig. 8.9). The image is formed on a CCD (§5.6) and the individual image pixels can be seen; it is clear that the telescope is close to the limit of its resolving power.

Figure 8.9   A ground-based telescope image of Pluto and Charon.

8.4. Infrared Astronomy

The wavelength distribution of the electromagnetic radiation emitted by a body is dependent on its temperature (Fig. 6.1) and many cool astronomical bodies emit mostly in the infrared part of the spectrum. The range of wavelengths constituting the infrared is between 700 nm, the boundary of the red end of the visible spectrum, and about 1 mm, beyond which we are moving into the realm of radio waves. The observational requirements over this range vary a great deal.

Although there is a sharp division in our perception of electromagnetic radiation between the visible and the neighbouring infrared and ultraviolet regions — we either see it or we do not — for telescopes there is no such division. In the near infrared and near ultraviolet, mirrors reflect, eyepieces transmit and CCD detectors detect, so images are formed in just the same way as they are for optical images. However, as the wavelength increases so there arise new requirements on the instrumentation. In the far infrared the objects being imaged have temperatures as low as 50 K or even lower, so it is necessary to cool the instrument, especially the detector, otherwise the signal from the radiation emitted by the object will be swamped by the radiation emitted by the detector.

Another, and very significant, difficulty with infrared astronomy is the effect of the Earth's atmosphere. Water vapour and other gasses in the atmosphere absorb infrared radiation and the atmosphere itself is a source of such radiation. Detecting objects in the mid-or-far infrared is like trying to see a dimly lit object against a brilliantly lit background. To overcome this problem infrared telescopes are usually situated in arid parts of the world at high altitudes, so reducing the amount of atmosphere between them and the objects they are viewing. An example of such an installation is ESO's Atacama Large Millimeter/submillimetre Array (ALMA) situated at an altitude of 5,000 m in Chile. Even in the most favourable environments the adverse effect of water vapour is still present and observations with land-based astronomical telescopes are usually made at a limited range of wavelengths for which water vapour is transparent (Fig. 8.10).

Since they are free of atmospheric problems, space telescopes provide the best infrared observations, particularly in the far infrared. Three outstanding examples are NASA's Spitzer Space Telescope, NASA's Wide-field Infrared Survey Explorer and ESA's Herschel Space Observatory (HSO), which also has NASA participation. The last-named instrument has the distinction of having the largest

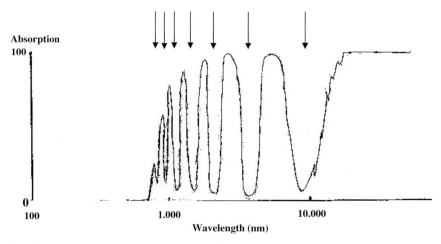

Figure 8.10   The transparency windows for water vapour are shown by the arrows.

mirror ever to be carried into space, 3.5 m in diameter. This mirror is made of silicon carbide, SiC, a synthetic ceramic material distinguished by having low density, high strength and low thermal conductivity, all ideal qualities for a space telescope. The HSO has a mass of 3.3 tonnes and orbits the Earth at a distance of 1.5×10^6 km with an orbital period of one year. It is always on the far side of the Earth away from the Sun and the coincidence of its orbital period around the Earth and that of the Earth around the Sun means that it is shielded from most of the Sun's radiation; it is at a special position relative to the Sun and Earth, called the *second Lagrangian point*, which is stable so that the Earth is always on the line between the Sun and HST. The detectors on the HSO are two bolometer arrays for the camera (§6.2.3) and photoconductor detectors in the spectrometer. These detectors are kept at a temperature below 2 K by liquid helium cooling; the telescope was launched in May 2009 with 2,000 litres of liquid helium, corresponding to a working lifetime of three years. The objectives of the HSO are primarily to study:

- Galaxy formation and evolution.
- Star formation and its relationship to the *interstellar medium*, the low-density material that occupies the space between stars.
- The chemical composition of planetary atmospheres.
- Molecular chemistry in all parts of the Universe.

An example of the power of the HSO is revealed in an image it produced of the Whirlpool Galaxy (M51), a spiral galaxy similar to the Milky Way. A conventional, but high quality, optical image is shown in Fig. 8.11(a). Figure 8.11(b) shows a false colour composite image taken by the HSO where a red image taken at 70 μm is combined with a green image taken at 100 μm and a blue image taken at 160 μm. The additional information given by the HSO image can clearly be seen.

With infrared imaging it is possible to see the cool clouds of gas and dust that are the birthplaces of star formation and stars being formed within them when they are at a low temperature, of the order of 100 K or so. An early source of high quality near infrared images was the Hubble Space Telescope (HST), launched by NASA

(a)   (b)

Figure 8.11   (a) An optical image of M51 (Todd Boroson (NOAO), AURA, NOAO, NSF). (b) An HSO image (ESA/NASA).

in 1990. There was an initial problem due to the 2.5 m mirror being incorrectly figured but this was corrected by fitting subsidiary optical components, installed during an Endeavour Space Shuttle mission at the end of 1993. Since then the HST has provided images of exceptional quality over a range of wavelengths from the ultraviolet through to the near infrared. Figure 8.12 shows a pair of images of the Trapezium Cluster, situated within a star-forming region, the Orion Nebula. The optical image (Fig. 8.12(a)) shows a fuzzy image with very little detailed structure. The same region in the infrared (Fig. 8.12(b)) shows many well-defined discrete sources, which are newly forming stars at low temperatures, which make them strong sources of infrared radiation but much weaker sources of visible radiation.

8.5. Adaptive Optics

The problem of atmospheric turbulence for ground-based telescopes means that, even with the largest mirrors, resolution is limited to about 1". The theoretical angular resolving power of a telescope, θ, is given by

$$\theta = \frac{1.22\lambda}{D}, \tag{8.2}$$

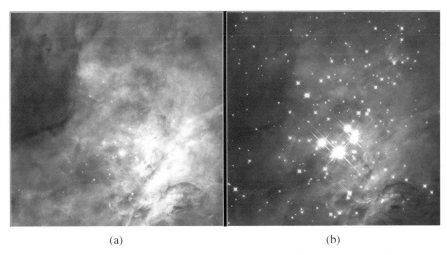

(a)   (b)

Figure 8.12   The Trapezium Cluster imaged with (a) visible light, and (b) infrared radiation (HST/NASA).

which is equivalent to Eq. (4.3) for θ small when $\sin\theta$ can be replaced by θ. For a large mirror, with $D = 8$ m and $\lambda = 550$ nm, the middle of the visual range, this gives $\theta = 8.4 \times 10^{-8}$ radians $= 0.017$".

The distortion of the incoming wavefront from a very distant point object, such as a star, away from its ideal planar form will be larger for larger telescopes. This is related to the sizes of turbulent cells in the atmosphere; the greater the size of the telescope the greater is the number of turbulent cells through which the light entering it has passed. As a telescope increases in diameter so its theoretical resolving power improves and the deterioration due to turbulence increases. For telescopes with mirrors less than about 1 m in diameter the effect of atmospheric turbulence is relatively small but for larger telescopes it becomes the overriding factor in determining the performance. The HST, free of atmospheric disturbance, can attain a resolving power of 0.1" that, although not quite as good as its theoretical resolving power of 0.055", is much better than that of any conventional ground-based telescope, whatever its size.

The effect of atmospheric turbulence is that light reaching the primary mirror does so with random small phase shifts at different

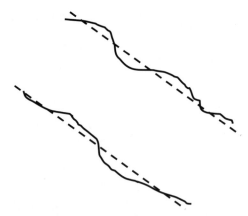

Figure 8.13   Two distorted wavefronts well separated in time. The dashed lines are undistorted wavefronts.

points of the mirror. Thus, thinking of a wavefront as a surface for which the phase of the light is everywhere the same, this is equivalent to having a distorted wavefront with the distortion varying with time. A two-dimensional representation of such wavefronts is shown in Fig. 8.13. A wavefront is a surface of constant phase and the changes of orientation of the wavefront surface as one moves from point to point on it can alternatively be considered as rays of light crossing different parts of the wavefront moving in slightly different directions.

There are great technical problems in launching large-aperture telescopes into orbit and the Herschel Space Observatory must be close to the limit of what can be done now — although it is feasible that a larger telescope could be assembled in space, much as has been done with the International Space Station. One advantage that a very large space telescope would have is that there would be no distorting effects of gravity so the mirror could be of lighter and less rigid construction. The large mirrors of conventional ground-based telescopes have to be very heavy and although glass or ceramic material can be chosen that is rigid and has a low thermal coefficient of expansion it is almost impossible to avoid some distortions during operation. The mirror of the Hale 5.1 m telescope at the Mount Palomar Observatory in California has a mass of 14.5 tonnes and is about the largest that can be made without gravitational distortion substantially affecting the image quality. From 1948 to 1993 the Hale telescope was the world's most powerful telescope. A slightly larger instrument, with a 6 m mirror, the BTA-6, was completed in 1976 and situated in the Caucasus Mountains of Southern Russia but it was dogged with technical difficulties, including cracks in its mirror, and so never lived up to its theoretical expectations.

8.5.1. *The Keck telescopes*

The first telescope to exceed the Hale in performance was the first of the two telescopes of the W. M. Keck Observatory completed in 1993 at a height of 4,145 m on Mauna Kea in Hawaii, the site of several observatories. These telescopes dealt with the problem of constructing a very large mirror by constructing them in sections. The total mirror is assembled from 36 hexagonal sections, arranged as shown in Fig. 8.14, giving an effective total mirror diameter of 10 m. The middle section is missing since, in any case, this would be blocked out by a secondary mirror (Fig. 8.5).

Each individual mirror segment has a maximum width of 1.8 m, a thickness of 7.5 cm and a mass of about 500 kg. The comparatively

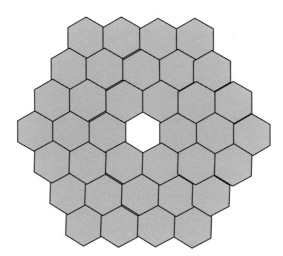

Figure 8.14 The 36 hexagonal mirror sections forming a Keck mirror.

thin sections were possible because each was so small, so the total mass of the Keck telescope mirror is not much more than that of the Hale telescope, albeit that it has about four times the surface area.

If this concept of having a segmented mirror is to work properly then it is important that the segments are properly aligned. To this end there are 168 position sensors mounted at the edge of the mirror segments that detect height difference between neighbouring sectors and can detect relative motions of the order of $10\,\mu$m. This information is analyzed by a computer and relayed to *actuators*, of which there are three per mirror section, whose motions vary the orientation and displacement independently for each mirror. However, the ability to move each mirror does not just enable the sections to create a perfect parabolic mirror surface but also enables a process of *adaptive optics* to be performed, which largely corrects for the distortions due to atmospheric turbulence.

Although atmospheric turbulence gives a wavefront distorted from the ideal shape — a plane for an object at infinity — over short sections of the wavefront the profile can be approximated as plane. A two-dimensional illustration of this is given in Fig. 8.15(a) in which the wavefront is approximated as six straight lines, each covering one-sixth of its total width. The straight lines are tilted and do not necessarily link end to end. If each of the straight lines is tilted and displaced to coincide to the ideal straight-line wavefront and then the corresponding actual wavefront in each section is tilted and displaced by the same amount then the match of the modified wavelength to the ideal one is shown in Fig. 8.15(b). The match is not perfect but much better than the original one.

Extending this two-dimensional example to three dimensions, if the 36 mirrors of a Keck telescope can be oriented and displaced by the correct amount the distortions of the plane wavefronts from distant objects by atmospheric turbulence can be much reduced. Finding the correct displacements and tilts requires the presence of a star fairly close to the object to be imaged. Some light from the field of view is deflected into a device, a wavefront analyser, which rapidly determines how to move each mirror to give the closest approximation to a point image for the star. The quality of the image obtained

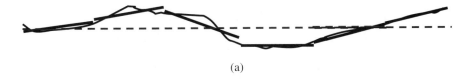

(a)

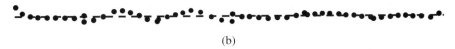

(b)

Figure 8.15   (a) A two-dimensional representation of an wavefront (thin full line), a straight line approximation to the wavefront in six sections (thick lines) and the ideal wavefront (dashed line). (b) The modified wavefront after adaptive optics (dotted line).

of an object using adaptive optics depends on how close it is to a suitable star and in some cases no close suitable star, i.e. one of sufficient brightness, can be found. An alternative procedure is to create a synthetic star by firing skywards a visible or ultraviolet laser beam that creates a small image by the scattering of light at a height of several kilometres. The best synthetic star is created with a sodium laser, the yellow light from which excites sodium atoms in the upper atmosphere at a height of about 90 km. With adaptive optics a Keck telescope can achieve an angular resolution between 0.03 and 0.06", compared to its theoretical value of 0.014".

There are now two similar telescopes at the Keck Observatory (Fig. 8.16) that can be used together as an interferometer. In this mode of operation they can achieve a resolution of 0.005".

8.5.2. *Flexible mirror systems*

Segmented mirrors are not the only way of producing a usable large mirror of reasonable mass while at the same time allowing the application of adaptive optics. The ESO 8.2 m VLT mirrors are cast in one piece and, if they were required to be rigid under gravitational forces, they would need to have a thickness of 1 m and a mass of more than 100 tonnes. Apart from the problem of casting the blank for such a monster it would need several thousand tonnes of structural support.

Figure 8.16   The Keck telescopes (NASA).

The solution is to make the mirror much thinner and deliberately flexible. Each VLT mirror is made of a vitro-ceramic material called Zerodur, is 17.5 cm thick and weighs 24 tonnes. The optical system and the form of the adaptive optics are shown in Fig. 8.17.

The adaptive optics is carried out by distorting the primary mirror with 150 actuators spread evenly over its lower surface and also by moving the 1.1 m secondary mirror. As for the Keck telescopes, the object scene must contain either a real star or a laser-simulated star and information from the wavefront sensor is passed to a computer that calculates the required motions of the actuators and of the secondary mirror to give a sharp stellar image. Figure 8.18 shows a VLT image of the Crab Nebula, the remnant of a supernova explosion observed by Chinese and Arab astronomers in 1054. At its centre, but not visible, is the Crab Pulsar, the neutron star that is the dense residue of the star that exploded. This image has about three times better resolution than could be obtained by the HST.

The advent of adaptive optics in conjunction with mirrors of much larger size left the Hale 5.1 m telescope far behind others in terms of performance. The primary mirror was in one piece and was far too thick to be distorted by actuators. However, adaptive optics

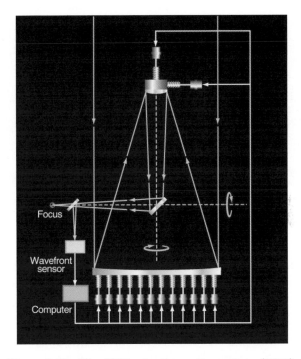

Figure 8.17   The VLT adaptive optics system (ESO).

Figure 8.18   A VLT image of the Crab Nebula.

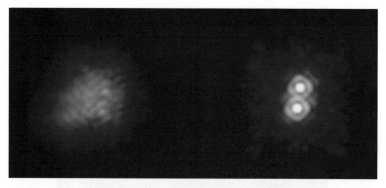

Figure 8.19　Hale telescope images of a binary star system without and with adaptive optics (C. Beichman and A. Tanner, JPL).

corrections have been applied at a flexible secondary mirror and the performance of the telescope has been enhanced enormously. As an example Fig. 8.19 shows the comparison between images taken without and with adaptive optics. The image with the adaptive optics clearly shows a binary star system; the stars have an angular separation of 0.3". Knowing that what is being imaged is a binary star, the image without adaptive optics just gives an indistinct indication of the presence of two stars.

An important discovery of the recent past is that of exoplanets, planets around distant stars. The first exoplanets were detected indirectly by the way that they affected the motions of their parent stars; by the very nature of such observations the easiest planets to detect were those of large mass in close orbits. However, in 2004 the first direct image of an exoplanet was made and since then several have been imaged. In 2008 the Keck 10 m telescope and the Gemini North Observatory, both on Mauna Kea in Hawaii, produced images of three planets around the star HR 8799. In 2010, as a demonstration of the power of the refurbished Hale telescope, just by using a 1.5 m diameter region at the centre of the primary mirror, the image shown in Fig. 8.20 was obtained, clearly indicating the presence of three exoplanets. The image of the central star has been blocked out, otherwise its light would flood the field of view so making the exoplanets impossible to image.

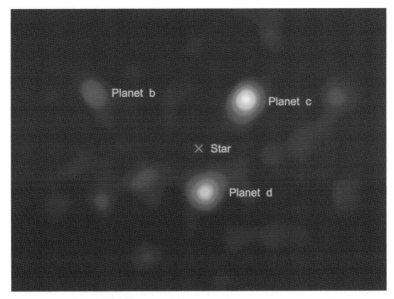

Figure 8.20   A Hale telescope image of three exoplanets around HR 8799 (NASA/JPL-Caltech/Palomar Observatory).

Given the introduction of adaptive optics, the emphasis in optical and near infrared astronomy has moved away from space-based back to ground-based telescopes. Ambitious projects are in train to build telescopes that will dwarf even the Keck telescopes. The European Extremely Large Telescope (E-ELT) project will have a segmented primary mirror with nearly 1,000 segments and a diameter of 42 m. It will be built in Chile and will possess advanced adaptive optics that will enable it to exploit its full resolving power — theoretically 0.003". With such an instrument the imaging of exoplanets, even for distant stars, should become routine.

Chapter 9

Imaging the Universe
with Longer Wavelengths

9.1. Observations in the Far Infrared

Figure 8.10, which shows the absorption by water vapour in the near infrared, gives only part of the story of the way that the atmosphere absorbs electromagnetic radiation. In Fig. 9.1 atmospheric absorption is shown over the whole range of the electromagnetic spectrum from γ-rays to radio waves. Wherever the absorption is 100% then ground observations are impossible and the only way to make observations at such wavelengths is to get above the atmosphere. These days that means mounting instruments in a spacecraft. It will be seen that, apart from some regions of partial transmission by the atmosphere, that is the only way to make observations in the far infrared.

The Big-Bang hypothesis of the origin of the Universe proposes that about 14.5 billion years ago an event occurred that created space, time and matter and also, but often forgotten, radiation. In the immediate aftermath of the Big Bang the temperature was extremely high with a corresponding high radiation density. High-energy photons could come together to create matter, in the form of particles plus their corresponding antiparticles, and those particles could later collide to reconvert mass back into energy. As the Universe expanded so the temperature fell, the photons became less energetic and the rates of conversion of energy into matter and matter into energy reduced. Finally matter and energy became decoupled and thereafter as the Universe expanded so the density of matter and the radiation

169

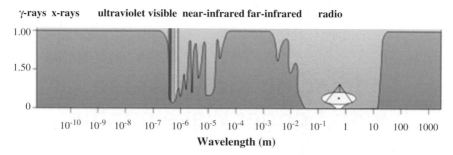

Figure 9.1   The total absorption of the atmosphere for electromagnetic radiation (NASA).

density both reduced with the consequence that the temperature, corresponding to the radiation density, also fell; from theoretical considerations it was expected that the present effective temperature of the Universe should be about 3 K. The way to test this prediction is to find the intensity distribution of the background radiation in the Universe from which the corresponding temperature (Fig. 6.1) can be found. From Wien's displacement law (Eq. 6.2) the peak of the curve at 3 K is at a wavelength of 1 mm, which is in the far-infrared region, where the atmosphere is opaque to radiation.

In 1989 NASA launched the COBE (COsmic Background Explorer) Earth satellite designed to test the theoretical predictions concerning cosmic background radiation. A Delta rocket launched it into a near-polar orbit such that, with suitable shielding, the instruments on board were constantly protected against exposure by radiation or particles coming from either the Sun or the Earth. The satellite (Fig. 9.2) carried three instruments that covered the range from the infrared to microwave radiation, the last named being in that part of the electromagnetic spectrum that spans the region where the far infrared merges into radio waves. The instruments were:

DIRBE — Diffuse InfraRed Background Experiment
FIRAS — Far InfraRed Absolute Spectrometer
and   DMR   — Differential Microwave Radiometer.

The first two of the instruments required liquid-helium cooling to which end the satellite carried 640 litres of liquid helium, sufficient

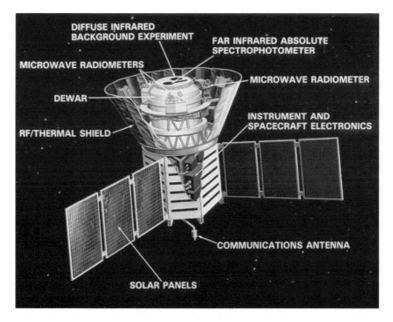

Figure 9.2   The COBE satellite (NASA).

to last for about ten months of the planned four-year lifetime of the mission.

The DIRBE instrument was designed to give absolute measurements of the radiation flux coming from different directions at ten specific wavelengths in the infrared region. These wavelengths are 1.25, 2.2, 3.5, 4.9, 12, 25, 60, 100, 140 and 240 μm. The radiation is filtered to the wavelength to be measured and a 32 Hz chopper (§6.2.5) exposes an absolute radiometer alternately to the incoming radiation and that coming from an internal helium-cooled surface. After the helium supply was exhausted DIRBE continued to operate but at much reduced precision.

FIRAS was a spectrometer, i.e. it measured the intensity of the radiation over the range of wavelengths from 100 to 1,000 μm, which spanned the region of maximum intensity of the expected intensity distribution. Its wavelength range overlapped that of DIRBE, which enabled checks to be made on the reliability of the results being found. Radiation was channelled into the instrument by a flared-horn antenna and was directed into a Michelson interferometer, a device that acted as a narrow-band filter. It compared the cooled-bolometer readings with those from a reference blackbody source (§6.1) within the instrument.

Unlike the other two instruments DMR was designed not so much to measure the absolute measure of radiation in the cosmos but rather to measure anisotropies, i.e. the variation of the radiation coming from different directions. It compared the temperatures in two directions 60° apart simultaneously; since both readings were affected equally by any thermal noise within DMR it was not necessary to have the detectors cooled. From the large number of readings taken it was possible to construct a map of the sky showing the variation of temperature with direction.

9.1.1. *COBE results*

The DIRBE results on absolute sky brightness gave some particularly interesting images of the galactic plane. The Milky Way is a spiral galaxy, somewhat similar to M51 (Fig. 8.11) with the Sun in a small spiral arm about 60% of the way out from the centre (28,000 light years) and close to the galactic plane. Figure 9.3 shows DIRBE images towards the centre of the galaxy taken at the ten wavelengths of its operation. The images clearly show the central bulge of the galaxy.

DIRBE galactic plane maps

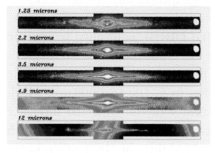 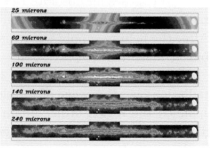

Figure 9.3   Images of the galactic plane at ten wavelengths from DIRBE.

The results from FIRAS confirmed with great precision that the radiation coming from the cosmos closely followed a Planck blackbody curve corresponding to a temperature of $2.726 \pm 0.010\,\mathrm{K}$. Over the range of wavelengths spanning the radiation peak the maximum deviation from the blackbody form was less than 0.03%.

The DMR results, shown in Fig. 9.4, were very revealing about the structure of the Universe in that it showed its lack of uniformity. Although the results at the different wavelengths are not entirely consistent they do show the same general pattern of variation of

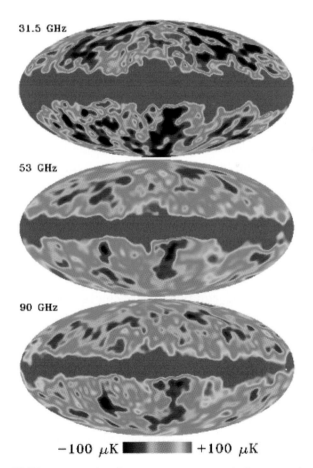

Figure 9.4   DMR images showing temperature variations as given at three different wavelengths.

temperature over a range $\pm 100\,\mu K$ from the mean. This result shows that the Universe is lumpy and probably was so from its origin. This would have been very important for its development as early lumpiness would promote gravitational instability. Large chunks of the Universe would have condensed to give regions of higher density that, by a process of hierarchical fragmentation would have given galaxy clusters, individual galaxies, stellar clusters and stars, eventually to produce planets.[1]

9.2. Radio Telescopes

It was realised quite early on that there would be a wide range of electromagnetic frequencies emitted by astronomical objects and that optical and near-optical frequencies were only a tiny part of the range. For ground-based astronomy — which was all that could be done before the space age was well advanced — observations were limited by the transparency of the atmosphere. However, while atmospheric absorption is a bane to astronomers it is a boon to life on Earth; but for the blocking of extreme ultraviolet radiation, and energetic radiation of even higher frequency, life on Earth would be impossible except in sheltered environments such as the depths of the oceans or within caverns.

In Fig. 9.1 it was seen that there is a wide window in the region of radio wavelengths, which gives the possibility of ground-based observations. It is a fortunate coincidence that radio waves, which require large observing equipment to achieve reasonable resolution because of their long wavelengths (Eq. 8.2), penetrate to ground level where large equipment can be constructed to detect them.

9.2.1. *The beginning of radio astronomy*

The undisputed father of radio astronomy was the American radio engineer Karl Guthe Jansky (1905–1950). After graduating in 1928 he was employed at the Bell Telephone Laboratories and was set the

---

[1]Further details of this process can be found in Woolfson, M.M. (2008) *Time, Space, Stars and Man: The Story of the Big Bang*, London, Imperial College Press.

Figure 9.5   Karl Jansky and his antenna array.

task of investigating sources of the static that interfered with the reception of radio waves. To this end he constructed a large antenna array (Fig. 9.5), about 30 m in diameter and 6 m tall on a rotating platform, designed to pick up radio waves at a frequency of 20.5 MHz (wavelength 14.6 m). Jansky was able to identify most of the static as due to thunderstorms but there was always a low-level hiss that could not be readily identified with any terrestrial source. The hiss varied in intensity with what at first appeared to be a 24-hour period and the initial interpretation of this was that it was due to emission from the Sun. However, as time went on it was clear that the period was actually 23 hours 56 minutes that, due to a combination of the Earth's orbit around the Sun and its axial spin, is the time for the Earth to spin through a complete rotation relative to the Universe at large. Jansky eventually found that this radio noise was coming from the direction of the centre of the Milky Way. These results were published in 1933 and attracted a great deal of favourable attention but, since the Bell Telephone Laboratory could see no practical application of the discovery, Jansky was not allowed to follow up his plans for further study of the phenomenon. There was little time for radio astronomy to develop to the point where its full importance was apparent before Jansky's early death at the age of 44, so he never received the accolade of a Nobel Prize that his discovery so richly deserved.

Following this early work, in 1937 an American radio engineer Grote Reber (1911–2002) built the first parabolic radio telescope, located in his back garden, an instrument 9 m in diameter with which he began to survey the sky looking for radio sources. The Second World War saw a temporary halt in development in this field and the next notable event was the building of a radio-frequency interferometer, observing 81.5 MHz radiation (wavelength 3.7 m), in the mid-1950s by the Cambridge astronomers Martin Ryle (1918–1984) and Anthony Hewish (b. 1924). This instrument consisted of four antennae at the corners of a quadrilateral 580 m in length in an east-west direction and 49 m north to south. With this they surveyed a portion of the heavens and produced the 2C catalogue (Second Cambridge Catalogue of Radio Sources) containing the positions and strengths of 1,936 individual sources. However, the results they obtained agreed poorly with the survey carried out with the Mills Cross Telescope, designed by the Australian astronomer Bernard Mills (b. 1920) and constructed close to Sydney. This device consisted of about 250 half-wave antennae situated along each of the arms of a cross some 450 m in both the east-west and north-south directions. The Australian telescope observed 85.5 MHz radiation, a similar wavelength to that used by Ryle and Hewish, but because of its design it gave much better resolution than the Cambridge instrument. Later the Cambridge interferometer was run at twice the frequency, 159 MHz, which improved the results and led to the 3C catalogue of radio sources.

A phenomenon observed by radio astronomers was that of *interplanetary scintillation* (IPS), analogous to the scintillation of stars, due to the diffraction of radio waves by the turbulent solar wind. To study this, and other phenomena, Hewish built a large array of 2,000 dipoles spread over an area of 1.8 hectares that swept a section of the heavens as the Earth spun on its axis. In 1967 his research student Jocelyn Bell Burnell (b. 1943) noticed an unusual signal, which she referred to as a *piece of scruff*, on the paper tape that recorded the days observations. It consisted of a burst of pulses at very regular intervals of 1.3 ms. After eliminating all possible terrestrial sources and noting that they were repeatedly observed from the same region of the heavens it was concluded that they were extra-terrestrial in origin. But what were they? One suggestion, briefly considered, was that they were some form of communication from an extra-terrestrial civilization, a story picked up by the popular press with the communicators labelled as *Little Green Men*. However, a year later the Austrian, later British, astrophysicist, Thomas Gold (1920–2004), correctly identified these objects, called *pulsars*, as *neutron stars*. A neutron star is the residue of a massive star, at the end of its main-sequence lifetime, after it has undergone a supernova explosion. It consists of tightly packed neutrons surrounded by an iron crust, with a mass typically somewhat greater than that of the Sun, usually about 1.4 solar masses, and a diameter of the order of 20 km. Because of their huge densities and small size they are able to spin very quickly without disruption. As they do so they emit a radio beam that sweeps the sky in the form of a cone. If this beam intercepts the Earth then a pulse is recorded and these occur at intervals equal to the period of the rotation. When Hewish received his Nobel Prize for Physics in 1974, which he did jointly with Ryle, the discovery of pulsars was one of the achievements that was quoted.

9.2.2. *Big-dish radio telescopes*

Grote Reber's parabolic dish, with a diameter of 9 m, seems impressively large, but for radio waves, even of wavelength as small as 1 m, its angular resolution was only about 8° — or 16 times the angular diameter of the Sun. For radio astronomy, to get reasonable resolution with a single-dish telescope, much larger dishes are needed.

Bernard Lovell (b. 1913), a British astronomer who worked on radar during the Second World War, conceived the idea of building a huge radio telescope. Following the war there was a great deal of surplus military equipment available that would help to keep down the cost. Lovell moved to the University of Manchester and began the construction of a 76.2 m diameter parabolic dish radio telescope

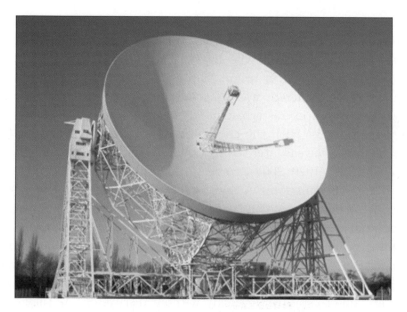

Figure 9.6   The Jodrell Bank 76 m radio telescope.

(Fig. 9.6) at Jodrell Bank, a quiet rural setting in Cheshire, some 30 km south-west of Manchester, which was relatively free of radio interference. As an example of the use of war-surplus equipment the steering mechanisms for the telescope had previously moved the gun turrets of British battleships.

The telescope was completed and became operational in 1957. It almost immediately established a role as a detector of space probes. In October 1957 the Soviet Union launched the first artificial satellite, Sputnik 1. It was a small object that did nothing but send out radio bleeps, but the Jodrell Bank telescope was the only one in the world capable of tracking the satellite's booster rocket. In the following years it tracked many other American and Soviet space vehicles; it picked up the transmissions of the Soviet Moon-lander, Luna 9, and was the first to distribute images of the Moon's surface taken by the lander. However, these well-publicized activities of the telescope took up little of its time and it began a fruitful scientific program, discovering and producing images of many radio sources in the Universe. At the time it was constructed the Lovell telescope was by far the most advanced instrument of its kind. It is still a major contributor to radio astronomy although there are two larger fully steerable radio telescopes — one of 100 m diameter in Effelsberg, Germany, and a slightly larger one at Green Bank, West Virginia in the USA.

There are clearly huge technical problems in building very large steerable-dish radio telescopes. A partial solution to this problem is to build them fixed to the ground where one can use terrain with a suitable curvature to reduce the amount of supporting structure required. Such an instrument, with a 305 m dish, has been constructed in Arecibo, Puerto Rico. It has a spherical, as distinct from parabolic, reflecting surface thus accepting some spherical aberration but this has the advantage that the receiving antenna can moved to receive radiation from different directions without any change in that aberration. In this way the telescope is effectively steerable up to 20° from the zenith and its receiving area sweeps out an annular ring in the sky as the Earth spins. An even larger semi-fixed telescope — the largest in the world — is the RATAN-600 completed in 1974 at Nizhny Arkhyz in Russia. It consists of 895 rectangular reflectors, each of dimensions 2×7.4 m, arranged in a circle of 576 m diameter, all pointing towards a central conical receiver (Fig. 9.7). It can be used over a wide range of wavelengths and when used with centimetric radio waves can achieve a resolution of 2".

9.2.3. *Radio interferometers*

From (Eq. 8.2) it can be seen that the combination of the long wavelength of radio waves and the practical limitations of building very large dishes limits the resolution that can be obtained with single radio telescopes. To overcome this problem the information from groups of well-separated radio telescopes is combined to produce an interferometer in which the information from different pairs of telescopes is combined by *aperture synthesis* (§7.8) to give a resolution corresponding to the largest separation of telescopes in the group. It was for the development of the technique of aperture synthesis that Martin Ryle was awarded his Nobel Prize for Physics in 1974.

Figure 9.7   Soviet Union stamp of 1987 commemorating the RATAN-600.

There are many groupings of radio telescopes forming interferometers worldwide. The Very Large Array (VLA) consists of 27 telescopes, each of diameter 25 m, situated on the Plains of San Augustin in New Mexico. They move on tracks forming the letter Y, each of the three arms of which has a length of 21 km. This gives 351 different baselines, a good basis for aperture synthesis. Since the distances involved are comparatively small the outputs from the telescopes can be combined by cable connections. A much larger array in terms of separation, established in the UK, is MERLIN (Multi Element Radio Linked Interferometric Network) with seven telescopes in various sites in England and Wales with a maximum baseline of 217 km. This operates at frequencies between 151 MHz and 24 GHz: at 5 GHz (wavelength 6 cm) the resolution achievable is 0.04", equivalent to what can be reached in optical wavelengths with the HST. Dwarfing both these systems is the VLBA (Very Large Baseline Array) linking sites in North America, Europe, China, South Africa and Puerto Rico. For a radio wavelength of 1 cm this interferometer can achieve a resolution of about 0.0003".

(a)   (b)

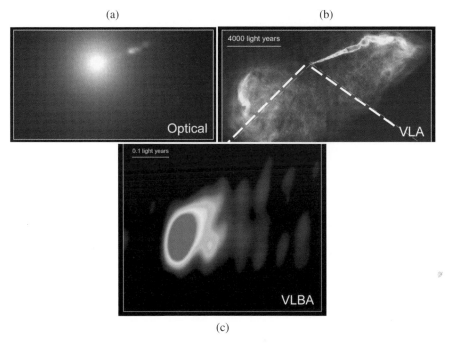

(c)

Figure 9.8   (a) An HST image of M87 (NASA). (b) An image of M87 produced by the VLA. (c). An image of a small region of image (b) produced by the VLBA. (Credits for (b) and (c): NASA, NRAO, NSF, John Biretta (STScl/JHU) and Associated Universities Inc.)

9.2.4. *Radio telescope images*

An interesting comparison between the types of information that can be gained from various kinds of observations is given by images of the galaxy Messier 87 (M87). In Fig. 9.8(a) an HST picture gives a rather bland image of the central part of the galaxy but there is the just the hint of a jet streaming out of the galaxy at 2 o'clock. The VLA image given in Fig. 9.8(b) shows a great deal of structure, particularly in the jet. The emission region of the jet is shown at much greater resolution in the VLBA image, Fig. 9.8(c). It is believed that the jet is being powered by a black hole situated at the centre of the galaxy.

Chapter 10

Imaging the Universe with Shorter Wavelengths

10.1. Some Aspects of Imaging in the Ultraviolet

Normal optical telescopes, with reflecting mirrors and CCD detectors, can quite comfortably produce images in the near ultraviolet. However, Fig. 9.1 shows that, quite close to the visible region, due to the ozone layer in the stratosphere, the atmosphere is opaque to ultraviolet light so ground-based telescopes cannot form images in the ultraviolet at wavelengths shorter than about 300 nm. Space-based telescopes, such as the HST, can go to shorter wavelengths, down to 115 nm, but at the shorter end of its range problems arise with the reflectivity of the mirrors.

Aluminium coated glass mirrors reflect extremely well at optical wavelengths and in the near infrared and near ultraviolet. However, for shorter ultraviolet wavelengths the reflectivity of aluminium falls and coatings of other materials reflect much better. In the ultraviolet range covered by the HST, coatings of magnesium fluoride or lithium fluoride reflect far better than aluminium, and the mirror of the HST has a 65 nm coating of aluminium overlaid with a 25 nm coating of magnesium fluoride, which both protects the aluminium below it and improves the reflectivity for ultraviolet radiation.

As an example of an ultraviolet image obtained by the HST, Fig. 10.1 shows a polar aurora on Jupiter. This is caused by the ionization and excitation of atoms in Jupiter's atmosphere, produced from collisions by particles, mostly protons, in the solar wind. Three of the bright dots in the image are the footprints of electric arcs

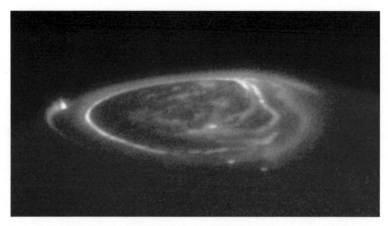

Figure 10.1   An ultraviolet Hubble Space Telescope image of a polar aurora on Jupiter (NASA/HST).

between Jupiter and its three nearest Galilean satellites, Io, Europa and Ganymede.

With its 2.5 m mirror the HST provides high-quality images for radiation ranging from the near infrared to ultraviolet. There are other space telescopes specially designed for operation in specific bands of the ultraviolet spectrum and three of these will now be described, selected to indicate the range of instruments available and the types of images and other information they provide.

10.1.1. *The International Ultraviolet Explorer*

The International Ultraviolet Explorer (IUE) was a joint project by the European Space Research Organization (ESRO, now replaced by ESA), NASA and the UK Science Research Council (SRC, now replaced by the Particle Physics and Astronomy Research Council, PPARC). In 1978 it was placed in a geosynchronous orbit, such that it orbits the Earth at the same rate as the Earth spins. This maintained its position over the mid-Atlantic and enabled it to be controlled and monitored from two control centres, one in the USA and the other in Spain. Because of the large energy required to propel a satellite into a geosynchronous orbit the weight of the IUE was an important consideration in its design. The primary mirror of the IUE, with the comparatively small diameter of 46 cm, was made of beryllium, a metal of low density — $1,850 \, \text{kg m}^{-3}$ — which also has a low thermal coefficient of expansion and high heat conductivity, meaning that the mirror's temperature remains uniform throughout, thus reducing distortion due to differential temperatures. The predicted lifetime of the IUE mission was five years but it actually operated successfully for nearly nineteen years.

The main purpose of the IUE was not to produce high-quality ultraviolet images; with its comparatively small primary mirror it could not have done so in any case. It contained an acquisition camera — the Fine Error Sensor (FES) — that produced an image with resolution about 8" and whose main function was to identify targets and then to track them with an accuracy of about 0.5". The radiation from the target was then fed via small mirrors, which concentrated the light, into two spectrographs, one working in the range 185 to 330 nm and the other in the range 115 to 200 nm. To optimize reflectivity the mirror coating for the shorter wavelengths was silicon carbide, SiC, and for the longer wavelengths lithium fluoride, LiF. The spectrographs found the intensity variation with wavelength of the radiation. As an example of the kind of information obtained by IUE we take the violent energetic star η-Cannae situated within the constellation Canna, about 8,000 light years from the Sun. This interesting star, with mass probably more than 100 times that of the Sun, undergoes occasional outbursts, emitting material and greatly increasing in brightness. After a particularly spectacular outburst in 1843 it was for a time the second brightest star in the heavens after Sirius. Figure 10.2(a) shows a Hubble picture of the star, in which can be seen material that has been violently ejected. In some way not understood, this material is a powerful laser source in the ultraviolet; the IUE spectrum in Fig. 10.2(b) shows two sharp peaks of laser emission.

What will be seen from Fig. 10.2 is that, in terms of visual impact, the image from Hubble far exceeds that of the output from the IUE. For all its lack of visual appeal to the layperson, in scientific terms

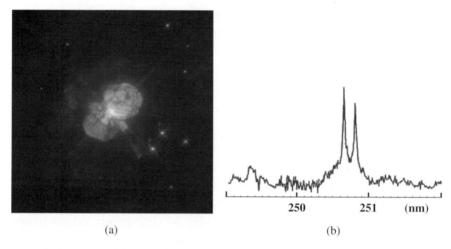

(a)   (b)

Figure 10.2   (a) An HST image of η-Cannae. (b) Ultraviolet output.

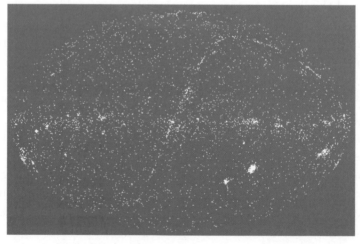

Figure 10.3   A celestial map showing the locations of objects observed by the IUE.

IUE was the most successful space satellite ever launched, often used in conjunction with other space instruments that operated at both longer and shorter wavelengths. It analyzed the ultraviolet light coming from thousands of objects varying in type from comets to galaxies. Figure 10.3 gives a celestial map showing their locations.

10.1.2. *The Extreme Ultraviolet Explorer*

In 1992 NASA launched the Extreme Ultraviolet Explorer (EUVE) satellite to examine objects at the very extreme ultraviolet wavelengths of 7 to 76 nm, the lower end of which range can be considered as longer wavelength x-rays. Its initial near-circular orbit, at a height of just over 500 km, decayed over time and the mission ended in early 2002 when the satellite plunged into the Pacific Ocean.

For the range of wavelengths used by the EUVE, and for shorter wavelengths, specular reflection does not occur from surfaces, except at grazing incidence where the radiation makes an angle of about 5° or less with the surface, the angle decreasing with decreasing wavelength. For this reason it is not possible to use normal parabolic mirrors to form an image but, instead, a special kind of lens system designed by the German physicist Hans Wolter (1911–1978) is used. There are three kinds of Wolter lens system but the simplest of them, the so-called Type-I, is shown in Fig. 10.4(a). The lens has two distinct components — both cylindrically symmetric surfaces, one with the profile of a portion of an ellipse and the other with a portion of a hyperbola as a cross section. The rays coming in parallel to the axis are reflected at grazing incidence first at the elliptical surface and then at the hyperbolic surface and come to a common focus on the axis. However, the requirement that the rays should fall on the surfaces at grazing incidence means that the cross section of the rays doing so on a single surface is limited. To increase the radiation collecting power of the lens a nested series of reflectors is formed, as shown in Fig. 10.4(b).

The EUVE had four Wolter-type telescopes of entrance diameter 40 cm and with gold-plated surfaces to increase reflectivity. Three of these telescopes were used to do an all-sky survey, designed to detect extreme-ultraviolet sources and to determine their brightness and temperature. These produced colour-coded overlapped images taken at 10, 20, 40 and 60 nm. The result of this survey is shown in Fig. 10.5 with the identity of the different types of source indicated by their colour.

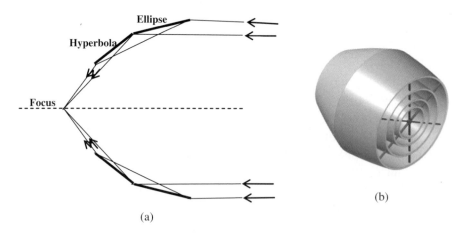

Figure 10.4   (a) A Wolter Type I simple lens. (b) A nested group of reflectors.

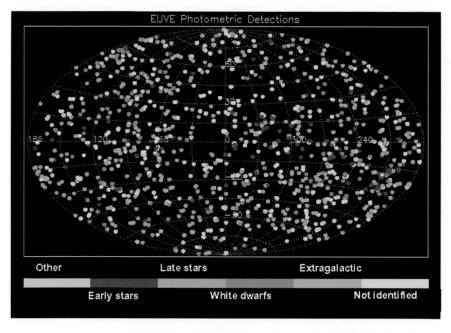

Figure 10.5   The EUVE all-sky survey.

The fourth telescope, with twice the focal length of the other three and ten times their sensitivity, was a deep survey and spectrometer (DS/S) instrument used to conduct a more detailed deep survey (DS) along the ecliptic at the same time as the all-sky survey, but just at 10 and 20 nm, and also function as a spectrometer (/S). One-half of the radiation passing through the telescope was sent to three spectrometers, instruments that analysed the radiation, giving the variation of intensity with wavelength. These spectrometers operated over three different bands of wavelength: short wavelengths (7–19 nm), medium wavelengths (14–38 nm) and long wavelengths (28–76 nm). From the data they collected it was possible to deduce the chemistry and the temperatures of the emitting objects.

Images obtained from the EUVE are of low resolution but they do indicate in a general way the nature of the extreme-ultraviolet radiation coming from different parts of the object. In 1996 the bright comet Hyakutake was observed, which passed comparatively close to the Earth to within 0.1 astronomical units.[1] It was the first comet from which extreme-ultraviolet and x-radiation was detected, although it has been subsequently discovered coming from other comets. Figure 10.6 shows a EUVE image of the comet that indicates different distributions of ultraviolet radiation coming from different parts of it.

10.1.3. *The extreme ultraviolet imaging telescope*

At the end of 1995, in a joint NASA-ESA project, the Solar and Heliospheric Observatory (SOHO) was launched. It is always on the line connecting the Earth to the Sun at a point at a distance of 1.5×10^6 km; this is another of the Lagrangian points of stability already mentioned in §8.4, in this case one that always ensures a line of sight to the Sun.

---

[1] An astronomical unit (au) is the mean distance of the Earth from the Sun and is 1.496×10^8 km.

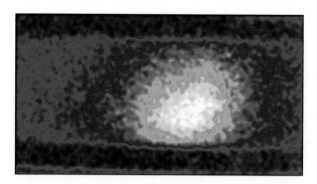

Figure 10.6   A EUVE image of the comet Hyakutake (NASA).

One of the instruments on SOHO was the Extreme Ultraviolet Imaging Telescope (EIT), designed to produce high-resolution images of the Sun at four wavelengths — 17.1, 19.5, 28.4 and 30.4 nm — corresponding to three wavelengths emitted by highly ionized iron and the other by ionized helium. Despite the very short wavelengths being used it is now possible, by improved vacuum-deposition technology, to produce mirrors that reflect in the normal way — but only at a particular wavelength. This is done by a multilayer coating on the mirror surface, consisting of alternate thin layers of a substance, such as silicon, that absorbs extreme-ultraviolet light very little and a strongly absorbing substance such as molybdenum. With about 100 layers of each substance, with the thickness of each of the order of 10 nm, a reflectivity of about 50% can be obtained at a wavelength determined by the thickness of the layers. With such mirrors one can construct reflecting telescopes of the types used for optical wavelengths.

In the EIT the mirror is divided into four quadrants with each quadrant coated differently to give reflectivity at the four chosen wavelengths. A moveable shutter can expose just one of the quadrants that then gives an image at the corresponding wavelength. This is recorded on a CCD detector with filters that pass the ultraviolet radiation while blocking out visible light that would otherwise swamp the CCDs. SOHO is constantly taking images of the Sun and transmitting them to Earth; Figure 10.7 shows a set of four images taken on 4 August 2010.

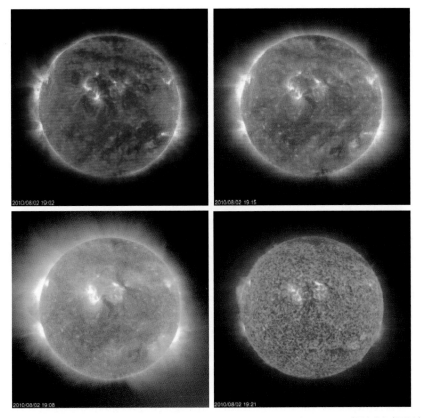

Figure 10.7 EIT images of the Sun at 17.1, 19.5, 28.4 and 30.4 nm (NASA/ESA).

10.2. X-ray Telescopes

Before the space age allowed satellites to be put into orbit, x-ray detecting instruments were being launched in the 1960s in rockets making brief observations before plunging back to Earth. The first flight, launched by American scientists in 1962, detected a bright source in the constellation Scorpius, which was appropriately named Scorpius X-1. The first satellites to be dedicated to x-ray astronomy, observing radiation in the range 0.01 to 10 nm, were launched in the 1970s and as time has passed so the sophistication of the observations has greatly increased. Observatories such as EXOSAT, launched by ESA in 1983, ROSAT, a German, American and British supported instrument launched in 1990 and Chandra, a NASA project and XMM-Newton an ESA project, both launched in 1999, have given extensive coverage of the heavens. The best of these instruments in terms of the collecting area for radiation is XMM-Newton, which uses 58 concentric Wolter-type mirrors to collect the radiation. However, because of the exceptionally high quality of its mirrors, shaped and polished to the highest precision and coated with iridium, the Chandra telescope assembly has the edge when it comes to resolution.

The Chandra optics consists of four nested Wolter reflecting surfaces (Fig. 10.4(b)). For radiation travelling parallel to the optic axis of the arrangement the grazing angle for reflection is smaller for the inner reflectors. Since the grazing angles at which reflection can occur is smaller for shorter wavelengths, for the very shortest wavelengths the outer reflectors are ineffective; the collecting area is $800\,cm^2$ for a wavelength of $5\,nm$ but only $400\,cm^2$ for a wavelength of $0.25\,nm$. The imaging takes place over a field of view $(30')^2$ in diameter. The telescope can be used as an imager or, if a spectrometer is placed in the focal plane, it can determine the x-ray spectrum.

The x-ray images from Chandra are often combined with other images taken with visible and/or infrared radiation since the composites are far more revealing than the separate images. An example of such a composite picture is shown in Fig. 10.8, which shows the remnants of a supernova observed by, amongst others, the famous German astronomer Johannes Kepler (1571–1630). The different images of Kepler's supernova remnant show very different aspects of the object but together they give a complete picture of the state of the constituent material.

An example of a Chandra image used on its own is the remarkable image of x-ray emission from the planet Mars shown in Fig. 10.9. Mars has a thin atmosphere, mostly consisting of carbon dioxide, CO_2. X-radiation from the Sun on colliding with an oxygen atom may remove one of the inner electrons situated closest to the nucleus. This gap is quickly filled by an outer electron that, in the process of moving from a higher to a lower energy state, releases energy that

---

[2]The symbol ' represents 'minutes of an arc'.

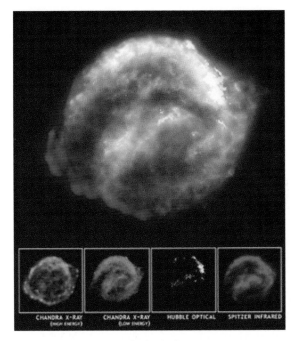

Figure 10.8   A composite image of Kepler's supernova remnant with contributions from high energy x-rays (blue), low energy x-rays (green), an HST optical image (yellow) and a Spitzer Space Telescope infrared image (red) (NASA/ESA/JHU/R. Sankrit & W. Blair).

appears in the form of an x-ray photon — a process known as x-ray fluorescence.

10.3. γ-ray Telescopes

Very-high-energy γ-radiation with wavelength below 0.01 nm, like x-radiation, does not penetrate the atmosphere and so must be detected by instruments in space. Early experiments were carried in high altitude balloons but the first γ-radiation detection experiment, carried on the Explorer XI satellite in 1961, which just picked up a few photons coming from all directions, was the beginning of significant advances in γ-ray astronomy.

Since γ-radiation cannot be reflected, no matter how small the grazing angle, other ways have to be found to determine the direction

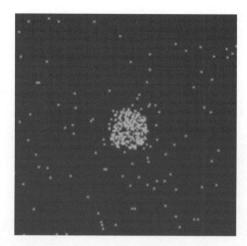

Figure 10.9   X-ray fluorescence from oxygen in the Martian atmosphere (NASA/CXC/MPE. K. Dennerl *et al.*).

from which they travel. For γ-rays the wavelengths are so small that they are difficult to comprehend so scientists tend to describe them by the energy of their photons instead. A NASA spacecraft, the Compton Gamma Ray Observatory (CGRO) was launched in 1991 and operated for nine years, collecting data over the enormous energy range[3] of 20 keV (wavelength 6×10^{-11} m) to 30 GeV (wavelength $(4 \times 10^{-17}$ m). It contained four instruments and we shall just consider one of them, the Energetic Gamma Ray Experiment Telescope (EGRET), to illustrate the kinds of techniques that are used to produce images with γ-rays in the energy range 20 MeV m to 30 GeV.

EGRET is an example of a pair telescope that depends on the phenomenon of *pair production*. When a very energetic photon interacts with an atomic nucleus its energy can be transformed into the mass and motion of a pair of particles, a negatively charged electron and its antiparticle, a positively charged positron. In order for energy to be converted into mass in this way the energy of the photon, according to Einstein's mass-energy equivalence equation, must at least equal $2m_ec^2$, where m_e is the rest mass of each particle

---

[3] 1 eV (electron volt) is an energy of 1.602×10^{-19} J. 1 keV is 1,000 eV, 1 MeV is 10^6 eV and 1 GeV is 10^9 eV.

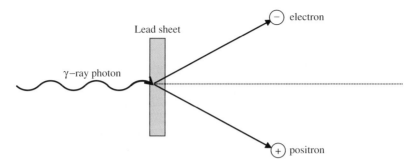

Figure 10.10   The process of pair production.

and c the speed of light. For an electron-positron pair this energy is 1.022 MeV, comfortably less than the energy range within which EGRET operates. Since the process is most effective for interactions with heavy atomic nuclei, lead is often the material of choice to give pair production. The process is illustrated in Fig. 10.10. A critical feature is that the line bisecting the paths of the two particles indicates the direction of motion of the γ-ray photon.

The standard device for detecting arriving γ-rays and determining the paths of the electrons and positrons produced by pair production consists of alternate layers of interaction layers — normally thin sheets of lead — and tracking layers that locate the positions of the particle passing through them (Fig. 10.11). The successive positions indicate the tracks of the two particles from which the direction of the incident γ-ray can be calculated. Having several interaction layers increases the probability that a pair-production process will take place. The energy of the incident γ-ray can be estimated either from the angle between the electron and positron tracks (which is energy dependent) or by placing at the bottom of the layered system a device that indicates the energies of charged particles. One such device is a scintillation detector in which energy from the charged particle is absorbed and transformed into flashes of light, the total intensity of which gives the energy of the absorbed particle.

A common form of tracking layer is the silicon strip detector. Silicon is a semiconductor material. Traces of impurity incorporated into silicon, so-called *doping*, can change its electrical characteristics.

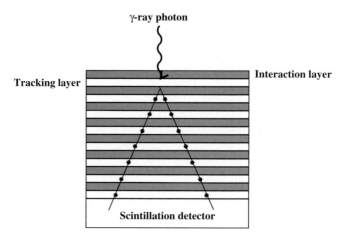

Figure 10.11   A schematic detector for pair production.

Silicon has four electrons in its outer valence shell by which it chemically binds with other elements — in the case of pure silicon to four other silicon atoms. If a small amount of phosphorus is added, which is a pentavalent element, then it becomes part of the silicon lattice but there is a spare electron, which is loosely bound to the lattice and will move to give a current under the influence of an electric field. This is called an *n-type semiconductor*, because it has charge carriers in the form of negatively charged electrons. Conversely, if boron, a trivalent element is added then it too becomes part of the silicon lattice but now there is a hole where in the pure silicon lattice there would be an electron. This hole behaves as though it were a particle with a positive charge equal in magnitude to that of the electron; under the influence of an electric field the hole will migrate through the lattice. Such material is called a *p-type semiconductor*.

A very simple type of semiconductor device is a p-n junction formed by slabs of p-type and n-type material in contact. If a potential difference is imposed on this device as shown in Fig. 10.12(a) then the holes will move towards the negative terminal, free electrons towards the positive terminal and, since there is no passage of charge across the interface between the slabs, no current will flow. However, if the potential difference is applied as in Fig. 10.12(b)

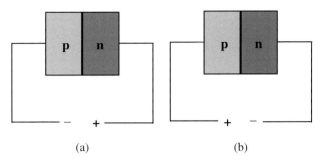

Figure 10.12   Semiconductor diode in (a) non-conducting configuration (reverse bias) and (b) conducting configuration (forward bias).

then electrons will be attracted towards the positive terminal, holes towards the negative terminal, charge will move across the interface and a current will flow. The device acts like a diode that lets current flow in one direction but not the other. When the potential difference is applied to get a current there is said to be a *forward bias*, otherwise a *reverse bias.*

The relationship between potential difference and current is shown in Fig. 10.13. It will be noticed that there is a small current on reverse bias; this is because the thermal agitation of material on both sides of the interface produces a small density of electrons and holes that migrate across the interface in reaction to the electric field. This small current increases with temperature because the number density of both electrons and holes increases with increased thermal agitation. At a high-enough reverse bias, at the breakdown voltage, the current suddenly increases. This is due to the thermally generated charges being accelerated to the point where, by collision they remove bonding electrons between silicon atoms and so generate new electron-hole pairs. These new charge carriers are themselves accelerated and in this way an avalanche of charge-pair generation occurs.

A reverse biased p-n junction can be used as a detector of ionizing radiation. If the bias is fairly high, but lower than the breakdown level, then the passage of a very energetic charged particle will cause large scale electron-hole pair production and the rapid migration of these charges across the barrier will create a pulse of current for the duration of the passage of the radiation.

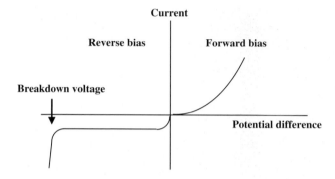

Figure 10.13   The voltage-current relationship for a p-n junction.

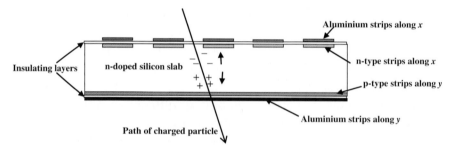

Figure 10.14   A schematic drawing of a silicon-strip tracking device.

A schematic drawing of a silicon-strip detector is shown in Fig. 10.14. On the top surface of the slab of slightly n-doped silicon, which is a fraction of a millimetre thick, there are fine parallel strips of heavily doped n-type silicon in what we define as the x-direction and on the bottom surface fine strips of heavily doped p-type silicon in the y-direction. Each strip has an overlying strip of aluminium, separated from it by a thin layer of insulating material. When a charged particle travels through the detector it ionizes silicon atoms. As previously described, charge will move through the device in the direction towards which they are directed by the reverse bias. They go towards the nearest strips giving them a positive or negative charge that then induces charges on the aluminium strips that are connected to read-out channels. The strips affected give the

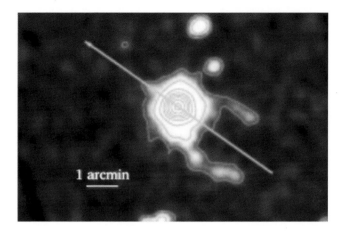

Figure 10.15 An image of the γ-ray source Geminga.

x and y coordinates of the particle. A positional precision of about 0.01 mm is attained with this device.

An interesting example of a γ-ray image is shown in Fig. 10.15 — a source known as Geminga, which in an Italian dialect spoken in Milan means 'it is not there'. The reason for this name is that this γ-ray source does not appear at other wavelengths. It is identified as a neutron star that is not detected as a pulsar because its radio beam does not sweep past the Earth.

Many γ-ray images show very little structure but they are important in indicating the occurrence of violent events, some happening in the far reaches of the Universe.

Chapter 11

Images of the Earth and Planets

11.1. Aerial Archaeology

The appreciation of a landscape depends very much on the scale at which it is seen. If you were in a warm climate by a pool of water, surrounded by trees and lush vegetation then you might think that you were in some fertile semi-tropical area. However, if you were transported a hundred metres vertically, so that your horizons were extended, then you might find that you were in an oasis surrounded by an extensive desert. So it is with detecting the remains of past human activity, a necessary process in archaeology. On the ground we see changes in the appearance of the soil here and there or slight bumps in the surface but such variations are common and normally are of no particular significance. It is only when we view the area from a considerable height that, sometimes, we see that these features are correlated in some way over a considerable area, perhaps forming unnatural shapes such as near-perfect circles, rectangles or long straight lines. Then we may suspect, or even be certain, that what we are seeing are the vestigial signs of some ancient human activity and know that digging in the area might reveal information about the past.

Aerial archaeology had its origin in the period before heavier-than-air flight had become a reasonably reliable form of transport. In 1906 an army balloonist in the UK took aerial photographs of Stonehenge and, in so doing, revealed features of the site not previously known. By the time of the Great War (1914–1918) frequent flights by aircraft were taking place and the combatants were making reconnaissance flights that included taking aerial photographs. An observer in the Royal Flying Corps, O.G.S. Crawford (1886–1957), realized the potential of aerial photography for archaeology and in the 1920s used RAF training flights to take photographs specifically for archaeological purposes. Crawford became the leading exponent of aerial archaeology in the UK in the inter-war period, 1918–1939, and after 1945 his work was continued and developed by others. Even during the Second World War aerial photographs were being taken that could be exploited for archaeological purposes; RAF planes, equipped with specialist cameras for the prime purpose of constructing up-to-date detailed maps, photographed the whole of northern France — an area rich in archaeological sites.

When structures are built, collapse wholly or partially and then become buried, they can leave traces of their presence, seen in various ways on the surface. Since the building material will differ from the local soil both chemically and physically the trace elements and nutrients in the soil immediately above the remnants of the building may differ from those elsewhere and alternatively, or additionally, the soil drainage above the remnant may differ from that in the general locality. Both these effects can be seen in the vegetation either in terms of colour or extent of growth. This is particularly true where crops are being grown; seen from a great height such areas usually seem uniform and featureless so that slight variations, especially organized into geometric shapes, are easily seen. These *crop marks* are visible in Fig. 11.1 and show clearly the presence of an archaeological site.

Another way that archaeological remnants can be detected is when an area has been repeatedly ploughed over a very long period of time, which tends to bring some subsurface material up to the surface. This can give rise to *soil marks*, where the soil above a remnant region is slightly different in appearance, usually in colour, from that in other places.

Outstanding examples of soil marking are the *Nazca lines* in Peru. These were not produced accidentally by ploughing but were the deliberate creations of the Nazca culture that flourished in the

Figure 11.1   Prominent crop marks in a field in Poland (Dariusz Wach).

region of the Nazca desert from 400 to 650 BCE. The Nazca people created them by clearing the reddish pebbles that cover the desert region, which exposed lighter whitish soil beneath. Figure 11.2 shows a monkey figure formed in this way. There are several hundred such figures of a geometrical form and more than 70 human, animal, bird or fish figures. Their purpose is unknown but they are possibly of religious significance although other origins and reasons for producing them have been suggested — including that they were created by extra-terrestrial visitors! Since some of these figures are nearly 300 m in extent, they must have required considerable effort and organization to produce, especially as their exact nature cannot be seen from the ground.

Another form of ground variation that can be of archaeological significance is that of variations of height. Sometimes these are so obvious that they do not have to be discovered from the air — for example, the *tumuli*, in the form of small hillocks, raised up above ancient burial sites. At other times they are just rather insignificant slight undulations of the ground that, to an observer standing on or near them, would appear to have no particular

Figure 11.2   The Nazca monkey.

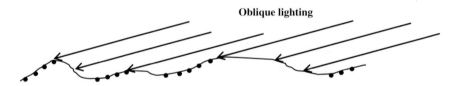

Figure 11.3   The areas indicated by black dots will be seen as shadows.

significance. From the air these variations of height can be seen very clearly under oblique illumination, such as occurs just after sunrise or just before sunset. With oblique lighting quite shallow features can throw large shadows, as illustrated in Fig. 11.3. These shadows stand out, starkly contrasted with the brilliantly lit illuminated areas.

These *shadow marks* can reveal very subtle features that would be completely missed by a ground observer. Photographs taken by satellites, as well as by normal aircraft, can reveal archaeological features on a very large scale. Figure 11.4 shows a region in Syria, the site of the ancient settlement of Nagar, taken by a Corona strategic reconnaissance satellite operated by the American Central Intelligence

Figure 11.4 The archaeological site of Nagar in Syria showing a road system indicated by shadow marks. The road indicated by the arrow shows shadows particularly well.

Agency. These images, now released to the general public, contain a great deal of information of archaeological interest. In this image shadow marks reveal a large network of roads that once existed around the site, indicating that, although Nagar began as a small Neolithic settlement about 6000 BCE, it must later have developed into an important centre of population.

There are other ways in which aerial photography can be used for archaeology. Very minor depressions, which are too shallow to give clear shadow marks, will sometimes contain small quantities of residual water when the surrounding soil is comparatively dry. If the temperature falls below zero then this water freezes and the shallow depression is clearly marked in white against a darker background. These *frost marks* will only occur in cold environments with particular patterns of precipitation. Another tool for aerial archaeology is infrared imaging (Chapter 6). Underground remnants can give small changes in surface temperature and a thermographic camera can pick these up.

11.2. Imaging Earth

The basic principle behind aerial archaeology is that, whereas from a limited viewpoint on the ground the significance of small local variations cannot be discerned, when we see a large area then small variations in appearance, producing unnatural shapes, indicate the presence of human activity.

Another situation where going to a larger scale of observation improves understanding is in the study of various phenomena associated with the Earth. By observations in one location it is possible to determine the current weather, the general form of the local landscape and the nature of the vegetation but not how this relates to the regional or global system. Once spacecraft observations became available we could see in a single view large enough areas to understand the complete structures of weather systems, geological features and patterns of vegetation. Observations and imaging from space satellites have revolutionized the way in which we can study the Earth and understand the forces that govern its behaviour.

We shall consider the impact of imaging from space in three main areas — weather, environmental science and mapping.

11.2.1. *Global weather*

Weather prediction has always been of importance to mankind. Gathering hay for winter fodder is best done when conditions are dry and the onset of seasonal rains can literally be a matter of life and death in generally arid areas of the world. Over time, patterns emerged that enabled short-term predictions to be made that, even if not perfect, had some statistical validity. Even today there are many who quote the well-known saying 'Red sky in the morning, shepherds warning; red sky at night, shepherds delight', which has an ancient origin. In the 1611 Authorized (King James) Version of the Bible one finds 'When it is evening ye say, It will be fair weather: for the sky is red. And in the morning, It will be foul weather today: for the sky is red and louring'. Earlier versions of the Bible give the same message, although in even more archaic language.

From the era of ancient civilizations, when cloud patterns wind direction and even astrology were used for prediction, through the ages there have been attempts to predict weather patterns but they were always hampered by the lack of rapid communication, which meant that the forecast had to depend just on local observations. An important indicator of local weather is the prevailing pressure and its rate of change. The domestic barometer measures the pressure and the expected weather is marked round a dial with high pressure indicating dry stable conditions and low pressure indicating rainy and squally conditions. These barometers are also provided with a pointer that can be set on the current pressure so that the direction of change of pressure can be found; falling pressure indicates deteriorating conditions and rising pressure indicates improving conditions.

The advent of fast communication systems, starting with the electric telegraph in 1835 and followed by radio communication, meant that information about atmospheric conditions over wide regions could be assembled in one place at one time. At the most elementary level, weather tends to travel with the prevailing wind so if the weather upwind is known, together with the wind speed, then predictions can be made over the shorter term. In the early part of the twentieth century theoretical meteorologists set up systems of mathematical equations based on fluid dynamics that could, in principle, forecast the state of the atmosphere on a global scale at future times, given the present state. However, it was not until the 1950s that computers made such calculations feasible and, even then, early computer-based forecasts for the day ahead were taking more than 24 hours to calculate that, although not practically useful, did serve to validate the computational method. With the advent of supercomputers long-term forecasts can now be made in hours and so are of practical use. What is needed for this calculation are worldwide data concerning the pressure, temperature, humidity and wind velocity at all levels of the atmosphere up to the stratosphere — a huge amount of information. To collect this information, all over the world there are observers, automatic weather stations, both on land and sea, taking data at ground level and, with weather balloons. In addition, commercial aircraft collect data from all levels of the atmosphere and transmit it to meteorological centres in many countries.

In 1961 an American meteorologist, Edward Lorenz (1917–2008) decided to repeat a weather-prediction calculation he had just completed. To save time he started the calculation with some intermediate results output by the first calculation as input for the new calculation. To his surprise this gave completely different results from the first run. It turned out that this was due to the input for the second run having a different number of significant figures from the numbers in the computer memory corresponding to that stage in the first run. An infinitesimal change in starting conditions led to an enormous change in outcome, meaning that predictions were only useful for short periods ahead. This discovery gave rise to a new branch of mathematics called *Chaos Theory*, which was shown to apply in many practical aspects of life. In 1969 Lorenz introduced the term *butterfly effect*, suggesting that a butterfly flapping its wings in California could drastically affect the weather all over the world in the longer term. The availability of super-fast computers now make possible the calculation of several trial forecasts made with various starting points covering the likely uncertainty in the measurements. These will give a variety of outcomes but one can see different probabilities for different outcomes in different regions. Thus a forecaster may say that, in a particular region, there is a 70% probability of rain, meaning that 70% of his trials indicated rain. Unfortunately, many members the general public do not understand statistics and, if it happened to turn out dry, they will form a poor opinion of the forecaster's ability.

The advent of weather satellites, starting in 1959, greatly improved weather-prediction capability. If they are launched in polar orbits then every region of the Earth is monitored every 12 hours. There are also geostationary satellites, launched in equatorial orbits at a height of approximately 36,000 km (orbital radius ~42,000 km), such that their orbital period is 24 hours, meaning that they constantly hover over the same region of the Earth. At these heights a few (theoretically three) such satellites can give complete Earth

Figure 11.5   Hurricane Katrina on 28 August 2005 (NASA).

coverage; in fact there are about a dozen of them with launching nations including, the USA, a European consortium, Russia, Japan, China and India.

Weather satellites can observe weather systems on a large scale and hence give predictions of major events, such as hurricanes or monsoon rains, affecting large areas. Figure 11.5 shows a satellite image of the 2005 Hurricane Katrina, which devastated areas of the southern United States and, in particular, caused catastrophic flooding and loss of life in New Orleans. This hurricane was tracked from the Bahamas where it had formed, across the Gulf of Mexico where it strengthened, then on to Florida and finally to the other coastal states on the Gulf of Mexico. Although predicting its path did not diminish its effects, it did enable warnings to be given that undoubtedly saved many lives.

The cameras on board satellites can take pictures of cloud formations on a global scale and also determine how they are moving. In the satellite images clouds appear as white; experienced meteorologists can recognize the different kinds of clouds and the types of weather systems that accompany them. The sea reflects little light and appears black and land appears in various shades of grey,

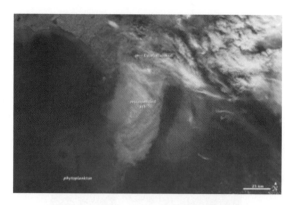

Figure 11.6   The ash cloud from Eyjafjallajökull, May 2010 (NASA).

depending on the type of terrain and the vegetation that covers it. The satellites also carry infrared sensors that record the temperature of clouds, land and oceans. They are able to make images of forest fires, clouds of smoke from fires and volcanoes, and eruptions of other material from volcanoes. In May 2010 the Icelandic volcano, Eyjafjallajökull, erupted and for some weeks sent clouds of ash over the North Atlantic and large parts of Europe. Since the ash could have adversely affected aircraft jet engines, even causing complete failure, air travel to, from and within parts of Europe was interrupted, causing large financial losses for many airline companies. The ash cloud (Fig. 11.6) was imaged and tracked by satellite observations.

The temperatures of land and sea and how they vary from place to place and from year to year have an important effect on the world climate that, in its turn, affects local weather conditions. An important meteorological event that occurs every three to seven years off the coast of Chile is called *El Niño*, Spanish for 'The Christ Child', since it tends to occur around Christmas time. It is a region of higher than normal pressure and higher than normal sea temperature, with corresponding lower than normal values in the Western Pacific region. This phenomenon, the origin of which is not understood, has effects in many parts of the world — the Americas, Africa, parts of East Asia and Australia — causing droughts in some areas

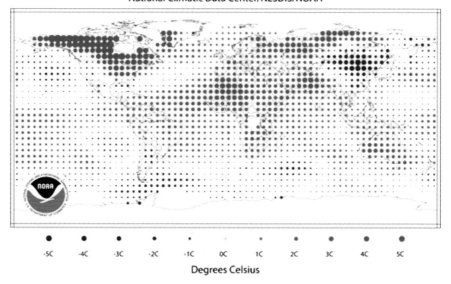

Figure 11.7   Temperature differences from an April 30-year average in April 2010.

and floods in others. Figure 11.7 shows a satellite-derived plot of the difference in worldwide April temperature in 2010 compared with the average April temperature in the 30 years from 1971 to 2000. Although some places are cooler and some warmer it is clear that the average temperature is higher overall — perhaps an indication of global warming. Such a plot would show an increase of sea temperature off Chile of about 0.5°C when an El Niŏo event was occurring.

11.2.2. *Imaging the Earth; environmental science*

Imaging the Earth from satellites began in 1972 with the launch by NASA of a satellite in a programme, originally called the Earth Resources Observation Satellites programme but later renamed *Landsat.* The purpose of the programme is to produce images of the

Earth's surface continuously over time to record changes that occur and to acquire information of relevance to agriculture, forestry, urban development, geology and changes in the Earth such as the spread of deserts or the destruction of rain forests. There have been seven satellites launched in the Landsat programme, although only six launches were successful. In 1979 the project was transferred to the National Oceanic and Atmospheric Administration (NOAA) and since 1985 has been run by a private organization, the Earth Observation Satellite Company (EOSAT). The Landsat images are made by a multi-spectral scanning device, covering the visible to infrared range of wavelengths, in which a scan is performed at particular wavelengths with a photomechanical device; the intensity information at any instant is transmitted along optic fibres to a photodiode and photomultiplier tubes. The image is then reconstructed from the stored information. Each scanned image covers a 185 km square of the Earth's surface at resolution of 57 m. A later Earth-observing satellite programme which began in 1986, *Satellite pour l'Observation de la Terre* (SPOT), a joint French/Belgian/Swedish collaboration, uses CCD recording of its images, with fixed-view imaging; each image covers an area of 3,600 km^2 with a resolution less than 5 m.

The aerial photographs in Fig. 11.8 give a shocking view of a man-made environmental disaster. The Aral Sea, an inland lake between Kazakhstan and Uzbekistan, was once the fourth largest lake in the world. In the 1960s these countries were part of the Soviet Union, which decided to divert the waters of two great rivers, the Amu Dar'ya and the Syr Dar'ya, to irrigate a desert region in Uzbekistan to grow cotton and other commercial crops. The project succeeded in that the crops grew but the effect on the Aral Sea was disastrous. It has now become three separate bodies of water that together have one-tenth of the original area and became so salty that fish could no longer survive, destroying once-thriving fishing and tourist industries. Exposed areas of seabed are covered in salt deposits that are blown by the wind over neighbouring areas, polluting them and making them sterile. The health of the local population has been adversely affected as has the local flora and fauna. There are projects in progress that are improving the situation but

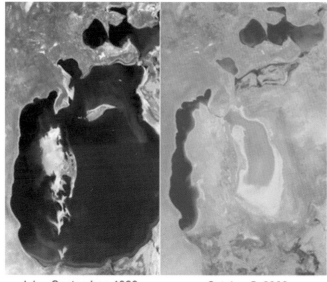

July - September, 1989   October 5, 2008

Figure 11.8   Aerial photographs of the Aral Sea (NASA).

these are slow-acting and unlikely to restore the original state of the environment.

Another example of the use of aerial photography in environmental science is Fig. 11.9 that shows a mysterious feature, known as the Richat Structure, some 50 km in diameter, in the Sahara Desert region of Mauritania. It does not have the characteristics of either a small asteroid impact, which would show shock features in the rock, or of the outcome of a volcanic eruption, which would give rise to igneous rock, both of which phenomena produce near-circular features. The probable cause is thought to be erosion of uplifted sedimentary rock — but why the feature should be so close to circular is not understood.

Earth imagery from space is now an essential tool of environmental science. With it scientists can monitor large-scale Earth-changing phenomena such as the advance of deserts, particularly affecting sub-Saharan Africa, changes in the ice sheets of the Arctic and of Antarctica and deforestation — changes that can alter both

Figure 11.9   The Richart structure in Mauritania (Landsat 7, USGS, NASA).

the climate and sea levels worldwide and affect the lives of people everywhere.

11.2.3. *Making maps*

The history of precise land-based mapmaking goes back to 1793 when a project started by Giovanni Domenico Cassini (1625–1712), an Italian, later French, astronomer, was completed by his children and grandchildren. This project produced the *Carte de Cassini*, a detailed map of the whole of France, the first of its kind ever to be produced. Before that time other maps had been made going back to the Romans and beyond. Many of them were picturesque and depended on imagination rather than objective observation and measurement; an outstanding example is the *Mappa Mundi* (Fig. 11.10), displayed in Hereford Cathedral, dating from about 1300 and the largest mediaeval map in existence (158×133 cm). It is drawn on vellum, has Jerusalem at its centre and apart from showing Great Britain, Europe and Africa (the last two incorrectly labelled) also shows the location of the Garden of Eden at the edge of the world.

Figure 11.10   The Mappa Mundi (Hereford Cathedral).

Rather more realistic and useful is the map of the world produced by Abraham Ortels (1525–1598), a Flemish cartographer. This is shown in Fig. 11.11; it was a practical map of the world as it was then known and served the needs of seafarers of the time.

In modern mapmaking the first step is to create a baseline, which is to measure the distance between two points, several kilometres apart with very great precision. The usual accuracy required is of the order one part in a million, corresponding to just 1 cm in 10 km. This was traditionally done using standard length chains or calibrated steel tapes, measuring section after section of the baseline and adding the results of the individual section measurements. The process was difficult and time consuming; nowadays it is done by Electronic Distance Measurement (EDM) which depends either on measuring the phase delay in a modulated microwave or laser beam transmitted from one end of the baseline and reflected back at the

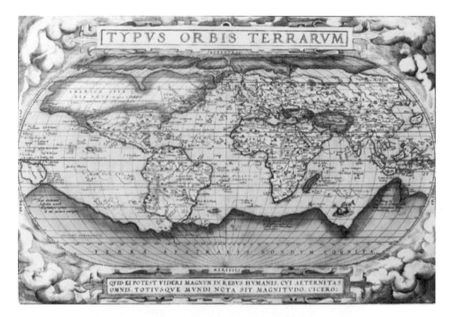

Figure 11.11   Abraham Ortels map of the world.

other or by the time for travelling the double distance by pulses of laser radiation.

Once a baseline is established a process known as triangulation can find the positions of a set of prominent points, called *trigonometric points* (trig-points), relative to the ends of the baseline. At the trig-point whose position is to be determined there is erected some kind of permanent marker, usually a concrete pillar that has on top a mounting base for a theodolite, a device for measuring angles with a rotating telescope; Figure 11.12 shows the trig-point at Wooton Warwen, near Stratford-upon-Avon in Warwickshire. A system of trig-points is established that cover the area to be mapped and their positions are progressively found relative to the baseline B_1B_2 by the triangulation process, illustrated in Fig. 11.13. A theodolite at position B_1 measures the angle AB_1B_2 and then with the theodolite at position B_2 the angle B_1B_2A is measured. These angles establish the position of A relative to the baseline. Now A and B_1 form an effective baseline for establishing the position of point C and measurements from A and B_2 can be used as a check. In this way, in principle, the

Figure 11.12   The trig-point at Wooton Warwen.

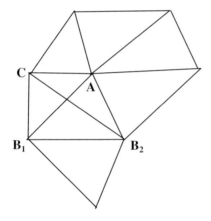

Figure 11.13   A system of trig-points with a baseline B_1B_2.

positions of all points can be found relative to the original baseline although in practice, as the distance from a baseline increases so cumulative errors make the positions less reliable. Theodolites can measure to an accuracy of less than 1" but this corresponds to an error of 5 cm over a distance of 10 km. To reduce accumulated errors several well-separated baselines are used and positions are checked for self-consistency of the complete triangulation system.

Having established a rigid framework of trig-points, two things are now necessary — to establish where these points are on the globe and where features such as roads, rivers, churches, bridges etc. are in relation to the trig-points. The position on the globe is found by astronomical observations of the Sun and stars, using the same theodolites as are used for triangulation. How the detail is filled in between the trig-points depends to some extent on the nature of the terrain. In favourable terrain, easy of access, the triangulation process can be repeated on a smaller scale to break each large triangle into much smaller ones. Then within the small triangles the positions of points that define the detail can be found by measurements from the apices of the small triangles. Less accurate measurements of angle can be made at this stage without introducing large errors in absolute position.

By measuring inclination angles as well as horizontal angles it is possible not only to locate the positions of places or objects in plan view but also find their heights. These may be indicated on a map as spot heights, for example at the positions of mountain peaks, or, more generally, as contour lines that join points of the same height. These are usually drawn at set intervals, perhaps 20 m or 50 m, and give a great deal of information from their general appearance. Where they are well separated the terrain is almost flat and where very close together there are steep slopes.

This kind of classical mapmaking has a long history. In 1790 the Board of Ordinance in the UK, which later became the Ordinance Survey, under the impetus of the threat of a possible invasion of Britain by Napoleon's army, began the task of making detailed maps of the country that would be useful to the military, starting with Kent, the most likely initial area of conflict. The work has been carried out continuously in the UK since then at various scales. In 1806 the British army began the *Great Trigonometric Survey of India*, setting up a triangulation system covering most of the country. In 1823 this task was supervised by Colonel George Everest who later became Surveyor General of India and whose name was given to highest mountain in the world, Mount Everest in the Himalayan mountain range. This triangulation became the basis of a detailed map of

India, in its time the best map outside some of the major countries of Europe.

Until the mid-twentieth century the process of mapmaking changed very little, although there were improvements of instrumentation and of forms of presentation of the final maps. Modern mapmaking now makes use of the latest technology. The *Global Positioning System* (GPS) based on signals from the available visible members of between 24 and 32 satellites, which is widely used by motorists for finding their destinations in unfamiliar territory, is capable both of accurately locating points on the Earths surface and of determining their heights. Additionally, aerial and satellite imaging is used accurately to portray the general form of terrain — coastlines, the routes of roads and rivers and the positions of mountains, for example. Imaging can be useful at many scales; Figure 11.14 shows an aerial photograph of the island of Seju, some 70×40 km in extent, situated off the western coast of South Korea. It is clear that this would provide the basis for a very detailed small-scale map. Satellite photographs can be used to produce large-scale maps; the aerial photograph shown in Fig. 11.15 of part of Washington DC could be a basis for a large-scale map of the city showing individual streets and buildings.

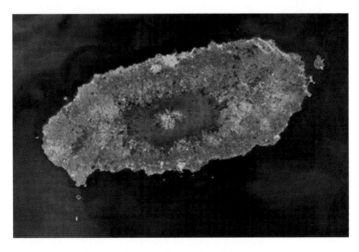

Figure 11.14   Seju Island (NASA).

Figure 11.15   An aerial view of part of Washington DC.

Figure 11.16   A Hubble Space telescope image of Mars (HST/NASA).

11.3. Images of Planets

Early Earth-based telescope images of planets showed them as disks and, in favourable situations, could reveal some large-scale features of their surfaces. With the launch of the HST and the development of adaptive optics much better images can be made; Figure 11.16 shows an image of Mars taken by the HST in June 2001 when the planet was close to the Earth, at a distance of 68 million kilometres. The polar caps can be clearly seen together with many detailed surface features, including some large dust storms. The resolution of this image is 16 km, which is remarkable if expressed in terms of the angular resolution — just 0.05".

Once artificial satellites were put into orbit around planets, images with a completely different scale of resolution became available. Figure 11.17 shows part of the surface of Mars imaged in November 2009 from a Viking Mars-orbiting satellite, launched in 1975. In the region shown there are many features that resemble dried-up riverbeds, which suggests that Mars once had an aqueous climate with rain and flowing rivers. The scale at the side indicates that the resolution is of order of a few hundred metres.

However, even Fig. 11.17 does not represent the limit of resolution that one can obtain of a planetary surface. Figure 11.18(a) gives a panoramic view of a Martian landscape made from a mosaic of 405 photographs taken by a vehicle, *Mars Exploration Rover Spirit*, which moves on the Martian surface. On a different scale

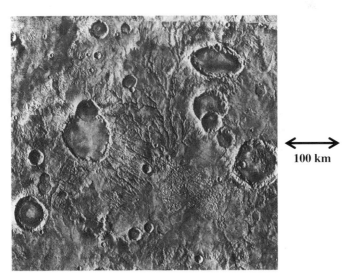

100 km

Figure 11.17   Part of the Martian surface imaged from a Viking spacecraft (NASA).

(a)

(b)

Figure 11.18   (a) A panoramic view of the Martian Surface. (b) Close-up of the surface showing small rocks and pebbles.

Fig. 11.18(b) shows part of the vehicle and small rocks and pebbles with resolution of a centimetre or less. The ringed area designated as 'Rock Garden', indicated by the arrow, is an area with a crust covering soft soil where some wheels of the Rover vehicle later became bogged down.

Images have been obtained from the solid surfaces of two other solar-system bodies. The Venera-13 probe, launched by the USSR in 1981, landed on the surface of Venus and survived for just over two hours under the harsh conditions at the planet's surface — a temperature of 457°C and a pressure of 89 atmospheres. During its brief lifetime it took a number of good quality pictures, including the near surface seen in Fig. 11.19.

Other remarkable images were made of the surface of the large Saturnian satellite, Titan, by the ESA Huygens probe, carried by NASA's Cassini spacecraft, which was launched in 1997 to orbit Saturn. The surface of Titan is covered with thick clouds of methane

Figure 11.19   Surface rocks imaged by Venera-13 (USSR/NASA).

Figure 11.20   The surface of Titan (NASA/ESA).

but could be imaged by altimetry radar and synthetic aperture radar
(§7.8). In 2004 the Huygens probe descended by parachute to the surface of the satellite, taking photographs on the way and also when
it reached the surface. Figure 11.20 shows lumps of water ice sitting

on a solid surface, the whole scene shrouded in an orange methane haze.

Since the beginning of the space age, vast numbers of images have been taken of every kind of solar-system object — planets, satellites, minor planets, asteroids and comets — over the whole range of the Solar System, from the Sun to the outer planets, and over a wide range of scales. Such images have answered many questions but also thrown up new questions as unexpected features of some of these bodies have been revealed.

Chapter 12

Images for Entertainment

12.1. Persistence of Vision

The act of seeing involves at its earliest stage the action of light on a visual pigment situated in a rod or cone. The light changes the conformation of the pigment molecule, which triggers a chain of events leading to the electrical impulses that travel through the optic nerve into the visual cortex. If a very brief, but bright, light impulse arrives on the retina the image will not be equally brief but will persist for a fraction of a second since the chemical processes operate over a finite time. This gives rise to the phenomenon known as *persistence of vision*, which has duration about 0.04 s. This means that when we look at a dynamic scene, one that is changing with time, what we see at any instant is a mixture of what the eye has received in the previous 0.04 s. Thus, if we look at a rapidly spinning child's top with pictures on its top surface we do not see rapidly rotating pictures but rather a blur, which is the average of the pictures at different distances from the spin axis.

Another aspect of the persistence of vision is that if we view a scene through a rotating shutter that briefly blocks out the view about 25 times per second then the eye is seeing the scene at all times because persistence of vision has maintained the viewing through the period of blockage. In such a circumstance we would be unaware of the presence of the shutter, apart from a diminution of intensity since some photons were being prevented from reaching the eye. If, alternatively, the view were interrupted ten times per second, with the blocked-out periods exceeding 0.04 seconds, then we would be aware of alternately seeing and not seeing the scene; there would be a distinct flicker effect, which would be annoying although we would still know what we were looking at.

Persistence of vision is of practical use in the operation of the *stroboscope*, a device that enables the motion of very rapidly moving machinery to be viewed. A stroboscope is a device that emits regular very intense, but very brief, flashes of light at a frequency that can be controlled. If the machine is operating at a rate of 6,000 revolutions per minute then this amounts to 100 revolutions per second. By having the machine in a dark environment with 100 stroboscope flashes per second, with each flash lasting a few, typically 10, microseconds, what is seen is the machine always at the same point of its cycle so that the moving parts seem to be stationary. The clarity of the image will depend on the duration of the flash; during a 10 μs flash the machine only moves through one-thousandth of its cycle so there will not be excessive blurring of the image. If the rate of flashing is slightly reduced then in each flash the part of the cycle seen will be slightly ahead of that seen in the previous flash and so the machine will be appear to be moving, but slowly. By slightly increasing the flash rate the appearance produced would be of the machine moving slowly with its motion reversed.

It will be seen from this description of persistence of vision that, although it may be thought of as some kind of degradation of the visual process because it limits time resolution, it is actually a very useful aspect of vision and one that we have learned to exploit.

12.2. Cinematography

The idea of being able to see moving images goes back a very long way. A crude moving picture can be produced by anyone with a book by drawing on the open corners of successive pages a figure, say a stick man, and slightly moving the limbs in going from each page to the next. If the pages are flicked through very rapidly then an impression of movement is obtained; persistence of vision ensures that there is continuity in the visual image. The basic requirement is that a sequence of images should be presented in rapid succession, with a small difference from one image to the next, representing a temporal sequence of views, and that there should be a blank period between successive images. If the blank period is short enough then it will be bridged by persistence of vision. For the purpose of general entertainment what is required is a moving image that can be seen by several people, preferably a large audience.

12.2.1. *Some early devices for moving images*

The advent of photography in the mid-nineteenth century gave the possibility of seeing realistic simulation of motion by presenting a succession of photographic images. A pioneer in this area was a colourful Englishman, whose life and career seem to relate to fiction rather than fact. He was born in Kingston-upon-Thames and given the name of Edward James Muggeridge (1830–1904) but he changed his family name twice, firstly to Muygridge and then to Muybridge and his first given name from Edward to Eadweard, the name of one of seven Saxon kings inscribed on the ancient coronation stone in Kingston, where these rulers were crowned. Muybridge spent much of his life in San Francisco, where he became an expert in the wet collodion process of photography (§5.2.4) and was well known for his artistic photography, particularly of architectural subjects and local landscapes.

A San Francisco business man, race-horse owner and ex-State Governor, Leland Stanford (the founder of Stanford University), was involved in a controversy as to whether or not there were times when all four legs of a racehorse were simultaneously off the ground. In 1877 he approached Muybridge to use photography to settle the issue. Maybridge eventually did so with the photograph shown in Fig. 12.1 that shows a horse clearly with four feet off the ground. It is interesting that some early paintings of racehorses show them with four feet off the ground, but with legs outstretched rather than bunched together under the horse as the photograph indicates.

Next Muybridge embarked on a project to take a series of photographs of a moving scene at very short time intervals so that the precise mode of movement could be accurately assessed. His first project was to produce a series of photographs of a galloping horse.

Figure 12.1   The Muybridge photograph of a galloping horse.

He set up a battery of cameras along a track, the shutters of which would be activated by strings placed across the track that would be broken as the horse moved past them. The exposures had to be short otherwise the individual photographs would have been blurred and, since photographic plates in those days were not very sensitive, the images were somewhat faint, but dense enough to be clearly seen and analysed for scientific studies. A sequence taken by Muybridge in 1882 of a galloping horse is shown in Fig. 12.2; the faint pictures have been enhanced by hand to show the silhouette of the horse and jockey.

Muybridge devised a piece of equipment, which he called a *zoopraxiscope*, to show these images briefly and sequentially to create the impression of continuous motion. The images, at first hand-drawn but later photographs, were arranged round around a disk as seen in Fig. 12.3. There was another disk of the same size with radial slits at positions corresponding to those of the images. The disks were mounted on a common axis and spun at the same rate but in opposite directions. At a position where one of the slits coincided with one of the images there was an optical system that projected the image onto a screen through the slit. For the brief time that the slit allowed the projection light to pass through, the image could be

Figure 12.2   Images produced by Muybridge of a horse in motion.

Figure 12.3   A zoopraxiscope image disk.

seen. The next time light passed through a slit would be when the next image could be projected. In this way a moving picture was seen on the screen. If it was a periodic motion, like the galloping action of a horse, then by photographing a complete cycle, a continuous long-term motion could be simulated. After his work on racehorses

Muybridge used photography and the zoopraxiscope to produce large numbers of moving pictures of many subjects — human and animal. However, despite this respectable pioneering work in creating moving pictures, to say that Muybridge led a colourful life is no exaggeration. In 1874 he shot dead his wife's lover and was subsequently acquitted by a San Francisco court on the grounds of 'justifiable homicide.'

12.2.2. *The beginning of cinematography*

Projecting images for entertainment has a long history. In the seventeenth century people were being entertained by the *magic lantern*, a device that could project an image drawn on a transparent glass plate, in those early days using a candle as the source of light. Later, lanterns were used to project glass-slide photographic transparencies to audiences — for example, to illustrate an expedition through darkest Africa, or some other nineteenth-century exploration. Then came the early experiments with displaying moving images, as illustrated by the zoopraxiscope and other Victorian devices of a similar kind. What could not be done in these early experiments was to show continuous motion of a non-periodic nature over a considerable period of time.

George Eastman (§5.2.4) made an important step towards the achievement of the goal of making continuous non-periodic moving pictures. He had perfected the process for producing dry photographic plates on a commercial scale but the first step towards moving pictures was his invention and production in 1884 of roll paper film. This was gelatine impregnated with photographic emulsion coated onto a long paper ribbon wound onto a spool. In 1888 this film, sufficient for 100 exposures, was provided ready loaded into a hand-held Kodak camera. Once the film was exposed the camera was returned to the company for the process of producing prints, the first stage consisting of stripping the gelatine off the paper to provide the negative. The next significant advance was soon afterwards, in 1889, when the company produced the first transparent celluloid roll film.

The individual who has the best claim to be called the father of modern cinematography is a Frenchman by the name of Louis Aimé Augustin Le Prince. He was born in 1842 but his date of death is unknown since he disappeared in mysterious circumstances on a train journey from Dijon to Paris in 1890. In 1866 he went to Leeds in England to join a brass foundry company owned by an English friend, married the friend's sister and then settled down there. He went to the USA, representing the company, in 1881; while there he became interested in the problem of producing moving pictures and invented and patented a device containing 16 lenses for doing so. Since each lens photographed the image from a different viewpoint, when the sequence of images was projected the resultant moving image jumped around. He returned to Leeds in 1887 and a year later patented a device using a single lens. This made use of Eastman's roll paper film. Essentially Le Prince's device repeatedly exposed the film, moved it forward one frame while it was shielded from light and then exposed it again; the rate at which this happened was between 12 and 20 frames per second. This gave a strip of negative images on the paper roll, which could be converted into positives but there was still the problem of projecting these to give the continuously moving image. Le Prince converted his individual images into glass slides and then experimented with various ways of projecting them to simulate smooth motion. Finally he mounted the slides on three fibre belts, with a mechanism that moved then alternately so that there was always an image being projected with minimal breaks, which reduced the flicker effect despite the low frame rate. The first motion image he produced in 1888 showed perambulating figures in a Leeds park; one of the stills is shown in Fig. 12.4.

Later, Le Prince used his camera to photograph the traffic and pedestrians on Leeds Bridge in the centre of the city. When these moving images were given a public showing in Leeds they were the first motion picture show to an audience in the modern sense.

The late 1880s and early 1890s was a very active period in the development of cinematography by many individuals in several countries. An English portrait photographer, William Friese-Green (1855–1921), often quoted as the 'inventor of cinematography'

Figure 12.4   A still from the first moving picture, produced by Louis Le Prince in 1888.

although he was marginally later than Le Prince, developed a camera in 1889 that was capable of taking 10 frames per second using perforated celluloid film. He demonstrated his moving images to the public in 1890 but his equipment was not very reliable and, because of the low frame rate, the moving images flickered a great deal and were rather jerky. Concurrently with Friese-Green's work, in the period between 1889 and 1892 the American inventor Thomas Edison (1847–1931) and his assistant William Dickson (1860–1935) were working on two devices — one for taking moving pictures, the *Kinetograph*, and another for viewing them, the *Kinetoscope*. The Kinetograph produced a sequence of images on perforated 35 mm film (Fig. 12.5(a)). The Kinetiscope (Fig. 12.5(b)) did not project an image but a viewer looked through an eyepiece to see the moving picture. The quality of the images was very good, but in the sense that it could only be seen by one individual it was inferior to systems that projected an image that could be seen by a large audience. Its importance, which explains the quality of the image, was that the system used a perforated film together with a high-speed shutter and became essentially the system used by the film industry thereafter before electronic means of storing images became available.

(a)   (b)

Figure 12.5   (a) A sequence of images produced on 35 mm perforated film by the Kinetograph. (b) The Dickson and Edison Kinetoscope, the forerunner of the modern cineprojector.

In the last decade of the nineteenth century the foundations were being laid for the film industry that would dominate the world of entertainment in the first half of the twentieth century. In 1892 a French inventor, Léon Bouly (1872–1932), designed the *cinématographe*, a device that could perform all the tasks of taking films, developing them and finally projecting them. He was unable to pay for the patent and in 1894 it was patented by the Lumière brothers (§5.4.1). With this equipment they made a series of short films, 10 of which, with durations from 38 to 49 seconds, they presented in 1895 in a public showing in Paris at the Salon Indien du Grand Café. The first of their short films was *La sortie des l'usines Lumière*, showing workers leaving the Lumière factory and the last, *La mer*, showing people bathing in the sea. So began a great industry. Through the work of the Lumière brothers, Paris was for many years the world centre of the film industry, until 1914 when the Great War (later known as the First World War) halted progress in Europe and thereafter Hollywood became dominant.

12.2.3. *The introduction of colour*

Early attempts to bring colour into films involved hand-tinting the original black-and-white films — which was a very tedious process although quite practical for some of the very short early films. The colouring schemes were usually very crude and it could be argued that, in some cases, the films were better left as they originally were. Another, and simpler, way of introducing colour was just to colour whole frames corresponding to particular scenes with a single colour to reflect the overriding lighting and action of the scene — yellow for candlelight, green for a country view, blue for a scene at night or red for a gory battle scene. The eminent Russian film-maker Sergei Eisenstein (1898–1948) produced films during the era of the Soviet Union during which time films, as well as other art forms, were required, where possible, to extol the virtues of the communist state. In his famous film *Battleship Potemkin*, made in 1925, which told the story of a mutiny on a Russian battleship in 1905, at one stage the mutinying sailors run up a flag, which was hand painted in red, the colour of the flag of the Soviet Union.

There was a strong incentive to extend the commercial development of colour photography for the still-picture market, which began in the first decade of the twentieth century (§5.4), into the production of motion films. One early entry into this market was an Anglo-French system called *Dufaycolour*, devised by the French inventor Louis Dufay (1874–1936) in the period 1907–1910. It was an additive colour process, similar in some ways to the Autochrome system, described in §5.4, except that the film base was coloured with red, green and blue dyes in a regular pattern (Fig. 12.6). The individual colour elements were very fine with about 150,000 of them per square centimetre.

The colour quality of Dufaycolour was good and amateur photographers used the camera film until the mid-1950s. However, its use for motion films was very limited, being used for only one full-length

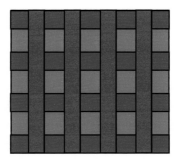

Figure 12.6 The pattern of red, green and blue dyes on a Dufaycolour film.

British film, *Sons of the Sea* in 1939, although several short films used the process. Despite being cheaper to produce than Technicolor it was eventually supplanted by Technicolor, the colour rendering of which was much more saturated and spectacular — although many would argue much less natural than Dufaycolour with its softer tones.

The term Technicolor is usually thought of as a single process that produced some of the Hollywood blockbusting films from the immediate pre-Second World War period until the 1950s, from the *Wizard of Oz* in 1937 to *The African Queen* (Fig. 12.7) in 1951. However, there were several Technicolor processes developed in the period from 1916 until they were replaced by better and cheaper processes from about 1960.

Figure 12.7 A poster showing Humphrey Bogart, the star of *The African Queen*.

The first Technicolor system, Process 1, was essentially an additive colour process, similar in principle to that used by Clerk Maxwell in 1861 (§5.3; Fig. 5.6). However, to simplify the process only two colours were used — red and cyan (green-blue) — so colour reproduction at the blue end of the spectrum was missing. Behind the lens of the cine-camera there was a beam-splitter that exposed two consecutive frames of a black-and-white film simultaneously, one through a red filter and the other through cyan. The projector had two lenses that projected the two frames simultaneously through corresponding red and cyan filters so that they were aligned on the screen. To produce the same number of visual images per unit time as a black-and-white film, and so avoid flicker and jerkiness, the camera and projector had to be run at twice the normal speed. Only one film was produced with this process, in 1917; it was difficult to project well as the projector needed frequent adjustment to align the images to avoid colour fringing on the screen.

From the experience with Process 1 it was decided that an additive colour system that required multiple projection was impractical. Again, if the film was broken up into small regions sensitive to different primary colours, as in Dufaycolour, then this inherently led to a loss of resolution. For this reason the Technicolor Process 2, developed in 1922, used subtractive colour, but still only using two dyes, as for Process 1. The initial step, producing pairs of positives of the same scene, one taken through a red filter and the other through cyan, was the same as that of Process 1. What was different was that the red- and cyan-filtered positives were then printed on the frames of separate filmstrips. The next step is chemically to remove the silver from the positive; the more silver that is removed the thinner is the effective layer of gelatine material that remains. The separate filmstrips are now dyed — the red-filtered strip with cyan-subtractive dye and the cyan-filtered strip with red-subtractive dye. When the original negatives were taken the red and cyan-filtered images were made the mirror images of each other. The final stage is to cement the two strips together in perfect registry and the Process-2 film is complete. The individual strips were one-half of the usual film thickness so the final product was of the normal thickness. Figure 12.8

illustrates how this process works, at all stages for a red part of the original scene but only showing the final cemented film for a green and yellow part of the scene.

For the red part of the scene the red-filtered positive image is more-or-less clear and the cyan-filtered image almost black. With the silver grains removed the effective width of the dye-absorbing film that remains is thicker for the red-filtered image than for the cyan-filtered. When the film is dyed and cemented together white light passing through the combination has wavelengths mostly removed at the green-blue end of the spectrum and the light passing through is dominantly red.

Figure 12.9 shows a scene from the first film produced by Process 2, *The Toll of the Sea*, in 1922. The colour rendering was far from perfect and, in particular, deep blues were impossible to reproduce — so film-makers using this process avoided scenes with much blue in them. There were other snags with Process 2. The cement sometimes gave way so that the two parts of the film separated in some places. Other problems were that there were two image-bearing surfaces to be scratched, so the films deteriorated more quickly than normal film, and the uneven thickness of the final film sometimes led to focusing problems.

In 1928 Technicolor developed Process 3, based on a different principle to those previously used — similar to the one used by Ducos du Hauron and described in §5.3.1. The first stage was similar to that for Process 2, exposing frames of two filmstrips simultaneously through red and cyan filters. Separate strips of film, coated with specially prepared light-sensitive gelatine, were then exposed to the red-filtered and cyan-filtered negative frames. When exposed to light the gelatine hardened so where the negative was clear the gelatine was hardest. A developer then washed away a quantity of gelatine in each part of the film, with the softer regions losing more gelatine. The residual gelatine was then dyed, with cyan-removing dye on the red-filtered frames and red-removing dye on the cyan-filtered frames. Hence the cyan-removing dye on the red-filtered strip was thickest where the scene was reddest and the red-removing dye of the cyan-filtered strip was thickest where the scene was closest in colour to

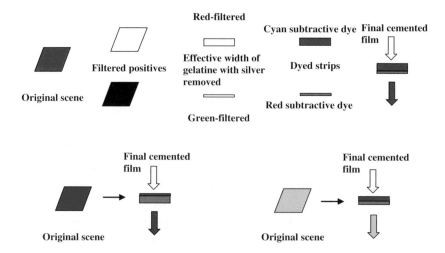

Figure 12.8   A representation of the Technicolor Process 2.

Figure 12.9   A scene from *The Toll of the Sea*.

cyan. Finally these dyes were transferred by contact onto a blank strip; the greater the thickness of the gelatine the more the amount of dye that was transferred.

Process 3 avoided most of the purely mechanical and optical problems of Process 2, except that a two-colour process could give only imperfect rendering of colour. This fault was effectively removed by Process 4, also involving dye-transfer, in which the final colours transferred were cyan, magenta and yellow to give perfect colour rendering over the complete range.

Technicolor was widely used until the mid-1950s when it was largely superseded by Eastman colour film in which all the colours were incorporated into a single strip of film (§5.4.2) and in which colour separation with filters was not necessary. Films were much easier to produce with this technology but the quality of the final product was no better than could be produced with the Technicolor Process 4. The Eastman films were also more prone to fading with time. Various other film-producing technologies were introduced and used from time to time — for example, Agfacolour from Germany — some of which were also used for amateur home movies. However, the advent of digital technology has now made such photographic processes largely redundant.

12.3. Television

The idea of transmitting an image over a considerable distance goes back to 1873 when the phenomenon of *photoconductivity* was discovered in the element selenium. In this process when light falls upon the material it becomes a better conductor of electricity, so this raised the possibility that light, the essential ingredient for image formation, could somehow be transformed into an electrical signal that could be transmitted over wires and then be reformed into an image.

12.3.1. *Mechanical scanning systems*

In 1884 a young German, Paul Nipkow (1860–1940) designed a process for converting an image into an electrical signal. At the heart of his equipment was a spinning disk containing square holes, arranged in a spiral pattern as shown in Fig. 12.10. If an image were projected onto the disk then the holes would scan across the picture in a series of curved strips — 16 in the case shown in the figure — that would completely cover the image. Behind the disk is a photocell, covering the whole area of the image, which converts the light passing

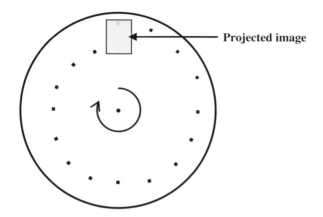

Figure 12.10   A Nipkow disk.

through a hole into a voltage that increases with the light intensity. This voltage can then be transmitted, in Nipkow's time by wire but later by a radio signal. At the reception end the voltage can be translated into a light intensity that is projected onto a rectangular area on a Nipkow disk synchronized with that being transmitting. In this way a spot of light passing through the disk scans over the rectangular area with intensity proportional to that passing through the hole at the transmitting end and the original image is reproduced. If the disks are rotated with sufficient speed then persistence of vision will give a smooth appearance to the final image and even moving images could be transmitted.

It is not known whether Nipkow ever made a working version of his equipment but it was certainly sound in principle. There were severe practical problems, in particular that the voltages generated by the photocell, which would need to be translated into light at the receiving end, were very small. With the development of the valve amplifier in 1907, which enabled these signal voltages to be increased, this difficulty could be overcome.

After the valve amplifier became available there were various developments of Nipkow's concept, with some modifications, until, in 1925, a Scottish inventor, John Logie Baird (1888–1946), gave the first demonstration of a working television system. By early 1926 this had been developed to a system that scanned quickly enough to give a reasonable moving picture with a 30-line scan that was just about capable of revealing details of a human face if it occupied the whole area of the image.

Baird began television broadcasts with his 30-line system, using a BBC transmitter, in 1929 and a year later regular broadcast programmes were being transmitted. In 1932 the BBC took over the transmissions and these continued on the mechanically scanned system until 1935. Then in 1936, by which time the number of mechanical scan lines had risen to 240, the service was switched from a mechanical system to one where the transmitted signal and the receivers were based on the use of electronics.

12.3.2. *Electronic scanning systems*

The development of electronics in the early part of the twentieth century stimulated many ideas for its practical application. As early as 1908 a Scottish electrical engineer, Alan Archibald Campbell-Swinton (1863–1930) suggested a method of using a cathode ray tube (§12.3.3) for both producing an electronic signal of an image and then converting that signal back into an image. However, this was a purely theoretical concept and early attempts to convert it into a practical system were unsuccessful.

In 1928 a young American inventor, Philo Taylor Farnsworth (1906–1971) demonstrated a working system for electronic television based on a device known as a *Image Dissector* that converted an image into an electric signal. An image, produced by an optical system, was projected onto a photocathode inside a high vacuum tube. The photocathode, a flat piece of photoelectric material, such as an alkali metal that emits photoelectrons when light falls on it, is maintained at a large negative potential relative to an anode within the same vacuum tube. Photoelectrons are emitted from each point of the photocathode at a rate proportional to the intensity of the image there and they travel straight towards the anode. The anode has a small aperture through which electrons can pass and behind the anode there is an electron detector of small receiving area. With this

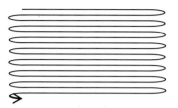

Figure 12.11   A raster scan.

arrangement the detector can only pick up the electron emission from a small region of the image on the photocathode and the current it generates will be proportional to the brightness of that region. The electron image is deflected by electric and magnetic fields so that it performs a raster motion, similar to that seen in Fig. 12.11. As the image moves so the detector generates a current, proportional to the brightness, from different parts of the image.

The Image Dissector has a low sensitivity to light because most of the photoelectrons being produced at any time do not reach the detector and hence do not contribute to the image signal. Several alternative approaches were tried to produce an electronic imaging system of greater sensitivity. A pioneer of this early work was the Russian, later American, engineer, Vladimir Zworykin (1888–1982) who invented an electronic imager called an *Iconoscope* in 1925. The central feature of this device was a *charge storage plate*, a portion of which is illustrated schematically in Fig. 12.12. On the conducting base plate there is deposited a layer of insulating material, e.g. aluminium oxide, on top of which is evaporated a very thin layer of photoelectric material, e.g. potassium hydride, in the form of tiny isolated patches. When an image falls on the plate each patch emits electrons and hence acquires a positive charge proportional to the amount of light it has received. The patch, separated from the conducting plate by the insulating layer, now constitutes a tiny capacitor. The plate is swept in raster mode by a fine electron beam which, as it strikes each patch, discharges it, thus giving an electrical output proportional to the intensity of light in the original image.

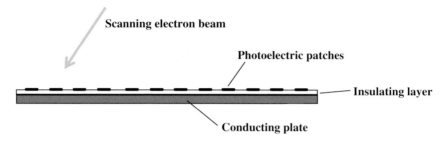

Figure 12.12   A schematic representation of an Iconoscope.

In principle this device produced a signal using the influence of all the light in the image but it turned out to give a very noisy signal. The reason for this is that when the scanning electron beam swept across a photoelectric patch it released further secondary electrons that were attracted to positively charged nearby patches and so neutralized some of their charge before the sweeping beam encountered them.

Ways of overcoming the noise problem of the iconoscope were devised, leading eventually to the *super-Emitron* produced by the British company EMI and the *Superikonoscop* by Telefunken in Germany. However, the basic principles involved are those described for the Image Dissector and Iconoscope.

12.3.3. *Television viewing with cathode ray tubes*

We have dealt with the process of converting an image scene into an electronic signal so now we must consider how this is reconverted into an image at the receiving end. The first way of doing this was by use of the cathode ray tube (CRT), invented in 1897 by the German physicist Karl Ferdinand Braun (1850–1918), who received the Nobel Prize for Physics in 1909, jointly with Italian scientist Guglielmo Marconi (1874–1957), for his contributions to wireless telegraphy. It was originally designed as a scientific instrument but has proved to be a device of great versatility, with uses such as looking at electrical

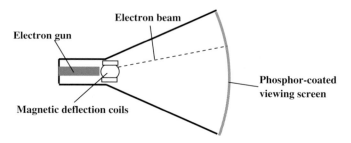

Figure 12.13   A schematic CRT.

waveforms, imaging radar targets and, our present interest, viewing television images. The basic design of a CRT used for television, all contained within a glass envelope, is illustrated in Fig. 12.13. The *electron gun* (§4.5) produces a fine beam of electrons of fairly uniform energy. The electrons from the gun are accelerated by passing them through a high-electric field, focused onto an anode by means of an electric-field lens and finally passed through a fine aperture to collimate the beam. A *control grid*, a wire-mesh arrangement near the cathode, the voltage on which affects the current passing through to the anode, modulates the intensity of the beam without changing the final energy of those electrons that do get through, which depends only on the potential difference between cathode and anode. This control-grid voltage is an amplified version of the signal voltage that indicates the varying light intensity of the original scene as it is scanned. In a CRT used for television the beam is deflected by magnetic coils, independently in two orthogonal directions, the fields, and therefore the extent of the deflection, depending on the current passing through them. By suitably varying the currents through the two sets of magnetic coils a raster motion of the electron beam can be produced, exactly matching that of the original imaging device.

On the viewing screen of the CRT there is a coating of a phosphor, a material in which atomic electrons are excited by the impinging electrons and then emit light of a particular wavelength when the atoms return to their ground state energy.

The number of lines in the raster scan dictates the definition of the image. The dominant system in the world, PAL, uses 625 lines

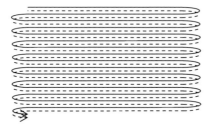

Figure 12.14   Interlaced scans as used in television imaging.

and generates 25 frames per second, which fits in with the 50 Hz mains electricity supplies in the using countries. In North America, parts of Asia and South America the NTSC system is used, based on 525 scan lines and 30 frames per second, the electricity supply being at 60 Hz. Finally, France, Eastern Europe and parts of the Middle East use the SECAM system based on 625 lines and 25 frames per second but differing from PAL in the way that it handles colour. All these systems use interlaced image fields (Fig. 12.14), alternately producing each half of a single frame, which helps to give smoothness to the image appearance and motion, so that the PAL system is actually making 50 raster scans per second.

Early in the history of television development, ideas were being advanced to introduce colour into the image. The first commercial colour broadcast, as distinct from private demonstrations, was made by the American CBS network in 1951. Colour television has now become the default mode of viewing and black-and-white television, if not quite extinct, is virtually so.

Colour television is based on the mixing by addition of the three primary colours, red, green and blue, in the way that Maxwell first produced a full-colour picture (§5.4) and it requires precision engineering of the very highest quality in the production of the associated CRTs. A colour CRT screen, examined closely with a magnifying glass while it is illuminated, is seen to be covered with a fine network of small circular patches of red, green and blue phosphors forming the pattern shown in Fig. 12.15(a). This pattern can be decomposed into tightly packed units consisting of one red, one green and one blue patch, arranged as in Fig. 12.15(b). If all three in

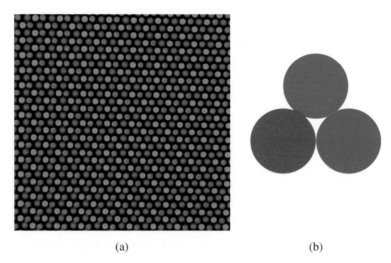

(a)   (b)

Figure 12.15   (a) The pattern of phosphors on a colour CRT screen. (b) A basic unit of the pattern.

a region of the CRT screen are glowing brightly at maximum intensity then the perceived colour is white or, if all are not illuminated, then the screen appears black. Brightly lit red and green and non-glowing blue patches give yellow and other combinations of patches illuminated to various degrees give all possible colour effects at a range of intensities.

The colour CRT contains three electron guns, one for each of the primary colours. When the television camera is producing the transmitted signal an optical arrangement separates the light from the scene into three separate channels which are then passed though red, green and blue filters before producing the transmitted signal. The television-set electronics separates out the three components, each of which modulates the intensity from one of the electron guns. Between the guns and the CRT screen there is inserted a *shadow mask*, a thin metal screen perforated with very fine holes. The directions of the beams from the red, green and blue guns, coming from different directions and passing through one of the holes, will only strike the corresponding red, green and blue patches of one of the units shown in Fig. 12.15(b). If the screen colour unit is as shown in Fig. 12.15(b) then the electron guns are arranged at the corners of a triangle. It is also possible to use a unit of three collinear phosphor patches, in which case the guns would be arranged in a line.

12.3.4. *Television viewing with liquid crystal displays*

Television screens based on CRTs, once the only available technology, are now being replaced by flat-screen television sets using either *liquid crystal* or *plasma* displays. Liquid crystal displays (LCDs) are ubiquitous in modern life — for example in various gadgets such as home computers, clocks, watches, calculators and telephones and also in car instrument panels and aircraft cockpit displays.

A liquid crystal is a material that is in a liquid form, so that it can flow, but in which the molecules take up ordered orientations, such as happens within a solid crystal. In normal liquids individual molecules are randomly oriented with respect to each other and remain so until the material changes its state to a solid. There are various forms of liquid crystal but here we shall just consider devices based on *nematic liquid crystals*. The word *nematic* comes from a Greek word meaning *threadlike* and the molecules in a nematic crystal are long and stringy. They tend to line up with each other, as shown in Fig. 12.16(a), and they have the property of affecting the polarization of the light (§3.4) that passes through them, which is that they only transmit light with electric vibrations in a direction parallel to the molecules.

A liquid-crystal display unit is illustrated in Fig. 12.17. The liquid crystal is sandwiched between two glass plates coated with a transparent conducting material, usually a metallic oxide, which act as electrodes. A thin layer of a polymer is then applied to this substrate and each polymer layer is rubbed in one direction by a cloth to create very fine microscopic parallel grooves. The liquid-crystal molecules adjacent to the plate are constrained to align themselves with the grooves. The grooves on the two plates are at right angles to each other and the combined effect of the constraints in direction at the plate surfaces and the tendency for neighbouring molecules to be aligned cause the molecules to gradually twist in going from one plate to another (Fig. 12.16(b)). On the outside of the glass plates

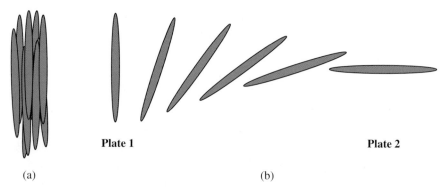

Plate 1   Plate 2

(a)   (b)

Figure 12.16   (a) The arrangement of neighbouring molecules in a nematic liquid crystal. (b) The variation of orientation of the molecules between the plates of an LCD.

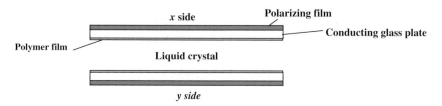

Figure 12.17   The structure of an LCD.

are two polarizing films that allow light to pass with the vibrations only in one direction; the allowed directions for the two plates are perpendicular to each other and the nematic-liquid molecules in contact with each plate are parallel to the allowed direction of vibration of the polarizing film on that side. For the purpose of explanation we will refer to the polarization directions of the two sides of the LCD as x and y. Light entering on one side is polarized in the x direction and moves into the liquid crystal where the molecules are so lined up that they allow free passage of the light. Now, the effect of the gradual twist in the direction of the molecules is to change the direction of vibration of the light as it passes through, always to be parallel with the molecules. By the time the light reaches the opposite plate the direction of vibration is in the y direction and so is able to pass freely through the LCD — it is transparent.

When a potential difference is applied across the LCD the molecules orient themselves towards the z direction, parallel to the field and perpendicular to the plates, and at a certain critical voltage they are completely oriented perpendicular to the plates. When this happens the twist in the vibration of the light passing through the liquid crystal no longer occurs. Light passing through the x polarizer arrives at the y side still vibrating in the x direction and hence cannot pass through the cell — it is opaque. As the voltage is increased up to the critical value the intensity of the transmitted light gradually reduces. It is clear that the LCDs do not produce any light of their own so when they are used for dynamic displays, such as a computer VDU (visual display unit) or a television screen they are back illuminated, usually by a fluorescent lamp behind the screen or sometimes by LEDs (light-emitting diodes).

The next question to address is how this technology enables an image to be formed; we restrict our initial discussion to forming a black-and-white image. An LCD image screen is made up of a large number of *pixels* (picture elements), each a tiny LCD, arranged on a rectangular grid. The image comes to the screen, after processing by the electronics of the viewing equipment, as a stream of different voltages that must be applied to the appropriate pixels in sequence to give the right shade of grey to give the image. An average display may have over two million pixels so it would be very difficult to have an individual circuit for each of them. The answer to this quandary is to utilize the rectangular grid structure of the display. The pixels in a single row are wired together and connected to the prevailing signal voltage source. The pixels in a single column are also wired together and go to different voltage sinks. When a signal voltage is received only one row is connected to it and only one column is connected to a sink, so only one of the pixels, that one common to the connected row and column, has a voltage applied across it. The usual procedure is to switch on one column and then run through the rows one at a time applying the prevailing signal voltage. The effect of this is to sweep through the pixels in raster fashion. There are two basic methods of carrying out this procedure, one known as *passive matrix* and another, and superior, one known as *active matrix* but

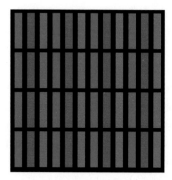

Figure 12.18   The structure of a colour LCD screen. Neighbouring red, green and blue elements form the sub-pixels of an image pixel.

they both pick out individual matrix elements by singling out its row and column as just described.

The introduction of colour to an LCD screen uses the same colour-addition process as is used for the CRT. Each pixel of the image consists of three elements, or sub-pixels, coloured respectively red, green and blue. The screen providing the colours is dyed in fine strips with the three colours giving a screen appearance as shown in Fig. 12.18. The method of exciting the pixels is exactly as described above although, for a given resolution, there will be three times as many pixels.

12.3.5. *Television viewing with plasma displays*

A plasma is often described as the 'fourth state of matter', additional to a gas, liquid and solid. A plasma is essentially an ionised gas in which atoms have lost electrons so that it consists of an intimate mixture of negatively charged electrons and positively charged ions. The household fluorescent lamp depends for its function on the properties of plasma. The lamp contains mercury vapour and an inert gas, such as argon at a low pressure — typically a few thousandth of an atmosphere. The cathode of the lamp is coated with a material that readily emits electrons when heated. An electrical potential applied across the lamp heats the cathode so producing electrons that are then accelerated, collide with inert-gas atoms and by ionizing them produces a plasma. The more plasma that is produced the higher is its conductivity and the higher is the current, to give even more plasma. To prevent the inevitable burnout that this would lead to, there is a safety device, a *ballast*, that limits the current flow. The mercury atoms in the tube are excited by collisions with electrons, meaning that some electrons of mercury atoms are pushed into higher energy states. These states are unstable and when the electron returns to its original state the energy released appears as an ultraviolet photon. The glass wall of the fluorescent lamp is coated with a phosphor that absorbs the ultraviolet light and re-emits the energy at one or more specific optical wavelengths.

A plasma screen is essentially a matrix of tiny fluorescent lamps and the production of such screens is a marvel of modern technology. The essential components of a plasma screen are illustrated in Fig. 12.19. Each 'lamp' is a tiny cell, filled mainly with neon and xenon with a very small amount of mercury.

The cells are connected separately in rows and columns by electrodes, those in the viewing direction being transparent. Above the transparent electrodes there is a transparent electrically insulating layer of a dielectric material and the whole arrangement is sandwiched between thin glass plates. At the base of each cell is a phosphor giving one of red, green or blue light; the arrangement of pixels and sub-pixels and the way they are triggered is similar to that used with LCD screens.

Plasma screens tend to be brighter than LCD screens and, because they emit light rather than just transmit it, they also offer better viewing at large angles. Very large screens such as are used in public displays are normally of the plasma type. They have the

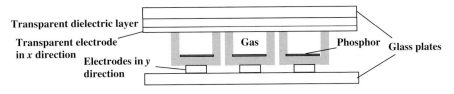

Figure 12.19   A schematic view of part of a plasma screen.

disadvantage of using more power than LCD screens but the difference is not large.

12.3.6. *Three dimensional television*

In §3.4 a method was described, using polarized light, for producing and viewing cinema films in three dimensions. This is clearly more difficult to apply to television and involves covering the screen with a fine chequerboard of elements in which neighbouring elements polarize light in orthogonal directions. The screen images for left and right eye fall on elements polarized in different directions and spectacles with polarizing lenses enable the two eyes to see separate images. A much simpler method is to project the left- and right-eye images simultaneously on the screen, respectively in red and cyan, as shown in Fig. 3.7 with the use of appropriate spectacles. While this would be a simpler solution it has the disadvantage that it would give a monochromatic image.

The basic requirement of stereoscopic imaging is to present different views of a scene to the two eyes. In all the examples given so far of how this can be done the two images have been presented simultaneously to the two eyes. However, there is an alternative, which is to alternately present the left eye with the left image and the right eye with the right image at a sufficient rate for persistence of vision to merge the two views. If each image is presented 25 times per second then a smoothness of motion similar to that given by cinema films will be obtained. This approach is now being exploited for television viewing.

The television camera has two lenses, imaging the scene from slightly different viewpoints. The two views are transmitted in interleaved fashion, each at a rate of 25 or 30 frames per second depending on the system being used. At the receiving end the viewer wears a special pair of battery-powered spectacles. Each eye looks through an LCD screen that alternatively goes transparent and opaque to let the eyes see the appropriate image on the screen when it is displayed. There is an infrared or radio linkage between the television set and spectacles that ensures synchronization of the viewing with the screen images. This system is well tried and tested and has been used by scientists for many years — for example to see three-dimensional images of the arrangement of atoms in a crystal.

Other ways are being developed of showing three-dimensional television pictures, for example using some kind of holographic process, but the main impediment to progress in this direction is probably the expense of both transmission and the television sets for reception.

Chapter 13

Detection and Imaging
with Sound and Vibrations

13.1. The Nature of Sound Waves

To most people the idea of imaging is restricted to that of forming
an image with light, although those of a more scientific bent might
extend the concept to imaging with electromagnetic radiation outside
the visible range. The essential property of electromagnetic radiation
that enables images to be formed is that it is a wave motion but
other kinds of wave motion also exist — notably sound waves. The
vibrations of electromagnetic waves are transverse — i.e. the electric
vibrations are perpendicular to the direction of motion. By contrast,
sound waves are longitudinal — the compressions and rarefactions
of the material in which the sound moves are along the direction of
motion. An impression of a sound wave is given in Fig. 13.1. Each
element of the material in which the sound wave moves is vibrating
to and fro in the direction of the wave motion and their combined
motions give a periodic variation of pressure and density that moves
along with the speed of the sound wave.

Just as it is possible to detect and image objects with electromagnetic radiation, e.g. with radar, so it is also possible to detect objects
and to produce images with sound. We shall see that mankind has
found several ways to use this aspect of sound for practical purposes.
However, our first description of using sound, just for detection, will
come from the natural world.

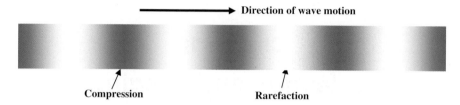

Direction of wave motion

Compression   Rarefaction

Figure 13.1   A longitudinal sound wave.

13.2. Animal Echolocation

A creature that is quite common, yet few people see, is the bat, the German name for which, *fledermaus* (flying mouse), describes its appearance quite well. It is indeed a mouse-like mammal, but with large leathery wings that make it an efficient and manoeuvrable flyer. There are more than 1,000 varieties of bat falling into two main categories — *megabats*, the diet of which mainly consists of vegetable matter such as fruit, and *microbats*, the diet of which is mainly insects but sometimes the blood of mammals. Bats, in general, are wrongfully thought of as evil and sinister creatures, mainly because of the reputation of blood-sucking *vampire bats*, associated with the fictional character Count Dracula, created by the Irish author Bram Stoker in the late nineteenth century. Indeed, the material injected by vampire bats into their animal blood source, which prevents the blood from clotting, is called *draculin*.

We are interested in the insect-eating microbats that have the problem that the majority of their insect prey are nocturnal and therefore that they have to be able to locate them in the dark. Despite the need to overcome this problem there are the compensating advantages that with night-time feeding there are few competitors around for their food supply and also fewer predators to worry about. To avoid predators during the day when they are inactive, microbats roost in dark and sheltered places such as caves or church belfries and in leaving these locations at night they need to negotiate many obstacles — another reason for being able to 'see in the dark'. A typical microbat is the *Big-eared Townsend bat*, shown in Fig. 13.2.

Figure 13.2   The Big-eared Townsend bat.

The solution to the bats' requirements, which has evolved by Darwinian selection, is the phenomenon of *echolocation*. This is essentially the process on which radar depends, except that it operates with sound rather than radio waves. The bat emits a sonic signal, then detects the target from the presence of an echo and finds the distance of the target from the delay between transmitting the signal and the return of the echo. Finally there must also be some method of determining the direction of the target. Then, like a radar-operating night fighter plane, the bat can close in on its target until it is close enough to snatch it from the air. The system sounds simple enough but it requires the bats to have many capabilities that all work in coordination.

The first requirement is to emit the sound, which is generated in the creature's larynx, and, as for radar, is sent in a pulsed form. The time interval between the emitted pulses of sound must be longer than the time between emission and reception if the distance of the target is to be precisely assessed without ambiguity. It will be seen from Fig. 13.2, as well as from the name of the animal, that this bat has very large ears that play the role of the receiving dish of a radar system. Even with these large sound receivers the maximum range of detection, which depends on the ability of the bat to detect the low-intensity returning echo, is just greater than 15 m, meaning that the total distance of travel of the sound is about 30 m. The emitted sounds have a very high intensity — 130 decibels being typical — which is the sound level of a jet plane flying overhead at a height of about 30 m. Indeed, the sound is so intense that, when it is emitted, special muscles close up the bat's ears to prevent the bat from being deafened by its own voice!

At a sound speed of 340 metres per second in air the return journey at a distance of 15 metres takes about 88 ms. When in a searching mode, looking for a potential target, the emission rate of pulses is in the range 10 to 20 per second, giving a maximum search range between 17 and 8.5 m. The frequency of the emitted sound within each pulse is somewhere in the range 14–100 kHz but normally about 50 kHz, well outside the range of the human ear that cannot hear sounds at a frequency above about 20 kHz. Scientists who study bats must detect their sound with the aid of special equipment that either shows a signal from their calls on a CRT or artificially slows them down to fall within the human acoustic frequency range.

A large part of the brain of a microbat consists of the auditory cortex that is devoted to analysing the acoustic information it receives. The direction in a lateral sense is determined from the difference of time between the sound arriving at the two ears; if the right ear receives the returning sound earlier then clearly the target is located towards the right. Given that bats are small animals and that the distance between their ears is about 5 cm, a distance that sound travels in about 150 μs, the time discrimination in the auditory cortex of a bat is remarkable. It can align its flight with the direction of the target by making the sound arrive simultaneously at its two ears. For up-and-down directional information the bat depends on the physical structure of the ear itself. The ear has several folds in its surface and sounds coming from above and below strike these folds at different parts of the ear, giving different sonic signatures that the auditory cortex interprets as vertical direction. Other items of information provided by the auditory cortex are the size of the target, estimated by the strength of the returning signal, and the speed and direction of the motion by a combination of transverse speed of travel and radial speed from Doppler-shift information.

Once the insect has been approximately located, the bat can start flying towards it; since it is a rapidly moving target the bat must make constant adjustments to its direction of flight. As it closes in, the requirements on the echolocation system change. The configuration of the bat relative to the insect is changing very rapidly so it is necessary to receive information more frequently. To enable this to happen the pulse frequency is increased although the condition must always be satisfied that the interval between pulses must be greater than the return time for the echo pulses. In the final approach stage the pulse frequency is about 200 per second, giving a maximum detection range of 85 cm — but by this time the insect's fate is sealed.

Other creatures that make extensive use of echolocation are the sea-mammals dolphins and whales, collectively known as *cetaceans*. They sometimes operate in murky environments, including deep under the sea or in muddy estuaries, and they can locate prey and obstructions by emitting sound pulses and detecting the echoes. Sound travels at more than 1,500 metres per second in seawater and can travel great distances — hundreds of kilometres — and still be detectable by a whale. The general principles that operate are similar to those that apply for bats. The emitted signals are in the form of clicks, one to five ms in duration, that are emitted at intervals of 35–50 ms, thus giving an echolocation range of up to about 38 m. The rate at which clicks are emitted is increased as the distance between the cetacean and the target reduces. The direction of the target is assessed by the difference in the strength of the signal received by the ears, which are different from human ears in that they are located internally and sound is transmitted to them through bones or fatty cavities in the jaw region. Different species of dolphin and whale have slightly different characteristics in the frequency of the sound in the emitted clicks and the range of sounds they emit. Beluga whales, sometimes called *sea canaries* because of the twittering noises they make, have been extensively studied. They emit frequencies in the range 40–120 kHz, well above the human auditory range; the actual frequency seems to depend on the ambient noise, the emitted frequency being as different as possible from that of the noise. Eleven different kinds of sound have been detected from Beluga whales, including clicks, whistles, squeals, chirps, mews, clucks, bell-like sounds and trills — all of which probably have some different information content for other Beluga whales. Dolphins emit whistles as well as clicks, which are probably a means of communication with each other — they are very intelligent creatures.

13.3. The Origin of Echolocation Devices

The ability of some animals to use echolocation has evolved to enable those creatures to occupy a particular niche in the environment and to give them an advantage in the battle for survival. The need to survive drives evolution and correspondingly, in the everyday affairs of mankind, there is a quote from Plato's *Republic* with a similar theme that declares that 'Necessity is the mother of invention'. In 1912, on its maiden voyage, the British liner *Titanic* struck an iceberg in the Atlantic Ocean and sank with a great loss of life. It was a moonless night and although the lookouts saw the iceberg it was too late for the ship to take effective evasive action. This raised the question of how the iceberg could have been detected long before it became visible and within a year echo-ranging patents had been filed both in Britain and Germany. In 1914 an American test of an echo-ranging system managed to detect the existence of an iceberg at a distance of 3 km, but not its bearing as the device had no directional discrimination.

During the First World War the threat of submarines to shipping injected new urgency into the search for a reliable method of detecting underwater objects. Initial British work was with the use of hydrophones — underwater microphones that could pick up the sounds of a submarine's propulsion system. French work concentrated on echolocation, modelled on that of the animal kingdom, in which a pulse of sound was emitted and the time for its return measured, together with some means of finding the bearing of the target. This was the basis of all future developments in this field. By the end of the First World War effective systems were developed, but too late to be actively deployed. In the early 1920s many British naval vessels had been equipped with an echolocation system known as *ASDIC*; the meaning of the acronym is not certain but may be 'Allied Submarine Detection Investigation Committee'.

In the Second World War ASDIC was extensively used in the Battle of the Atlantic, waged between German submarines and the Allied navies protecting merchant shipping. The objective of the surface ships was to detect the submarine and then to destroy it with depth charges that were dropped over the submarine's position and set to explode at its estimated depth. A weakness of ASDIC was that when the surface vessel approached very close to the position of the submarine, typically at about 300 m for an average-depth submarine, contact was lost and the submarine could then carry out various anti-detection manoeuvres, including sharply changing its direction of travel to take it out of the detection zone. Eventually, to counter the counter-measures, the ships used devices that projected the depth charges ahead of the ship so that they landed over the submarine while it was still being detected. Nevertheless, in the battle between submarines and surface vessels there was no decisive advantage to one side or the other. The critical weapon that eventually tipped the scales against the submarines was airborne radar flown by long-range aircraft, which could detect submarines on the surface at night when they were charging their batteries and illuminate them with powerful lights while attacking them, or even detect submerged submarines and attack them with depth charges or bombs.

13.4. Sonar

After the formation of NATO, the North Atlantic Treaty Organization, in 1949 the term ASDIC was dropped and now underwater echolocation is known by the American acronym SONAR (SOund Navigation And Ranging). There are two types of sonar, *active* and *passive*. For active sonar the requirements are basically the same as for radar, i.e.:

(i) *Formation of a narrow beam of pulsed emissions*
   The sonic pulses used in sonar are called *pings* because of the way they sound to the ear. The usual method of producing a

directed beam is by using an array of sound emitters coming from a single source; the general theory of the arrangement is described in §7.5.3 for the radar case. In sonar the emitter arrays can be one-, two- or three-dimensional, depending on the requirements of the system. Most ship-borne sonar has two-dimensional arrays but three-dimensional arrays, which give a much narrower beam, are becoming more common in both submarines and ships. The speed of sound in sea water is 1,560 metres per second, with some variation depending on conditions, so with a typical sound frequency in the range 1–2 kHz the wavelength is of order 1 m. The distance between the individual sound emitters is of order half the wavelength so it is clear that for a two-dimensional array of, say, 100 elements, a large naval vessel can comfortably accommodate the sonar equipment.

(ii) *Sweeping the beam over the angular range of interest*
Sweeping of the beam is achieved by varying the phase delays between neighbouring elements, the angle of the beam to the normal of the array then being given by (Eq. 7.4).

(iii) *Detecting the returning echo and determining the range of the target*
The returning echo is picked up by an underwater microphone and is then converted into an electric pulse by a transducer, a device that converts one type of energy into another — in this case the mechanical energy of sound into electrical energy (§14.6). This pulse is recorded on a CRT as a spot on a circular screen at a distance from the centre that gives the range, and in a radial direction that gives the bearing, of the target.

(iv) *Measuring Doppler-shift to find the radial speed of the target*
Unlike for radar, the Doppler shift in frequency in sonar can be an appreciable fraction of the emitted frequency. If the radial speed of an approaching target relative to the emitter is 18 km per hour, or $5 \, \text{ms}^{-1}$, then for a 1 kHz emitted beam the echo will have a frequency of 1,006.4 Hz, an easily measurable difference.

There are many other ways of applying active sonar in both military and civilian applications. Torpedoes can be fitted with sonar that detects the target and uses the returning signals to change direction and so follow a target taking evasive action. Modern fishing boats often use active sonar to detect shoals of fish, which enables boats to cast their nets only when there is some reasonable probability of making a substantial haul. We have already mentioned the close relationship of the applications of sonar and radar. One can have synthetic aperture sonar, similar to the SAR described in §7.8.1, with which one can image the seabed and detect objects, such as wrecks. Many wrecks of major large naval vessels and submarines sunk in both world wars have been discovered in this way.

The other form of underwater location of targets, passive sonar, is where the target itself is the source of sound. Individual people have characteristic *voiceprints*, which can be recognized by a spectrogram — the variation of frequencies with time in uttering a standard word or phrase. Similarly, a particular type of marine vessel will emit characteristic sounds dependant on its power source, onboard equipment and construction. Naval vessels engaged in passive sonar detection analyse the received signals in terms of their spectral composition; computers, provided with a library of sound patterns, can usually identify not only the type of ship emitting the sound but also provide a complete list of its characteristics — class within the type, tonnage, speed and weapon systems. A disadvantage of passive sonar is that, although the direction of the source can be found with a directional receiver, the range is not found directly. However, if two or more well-separated vessels are all receiving a signal from the target then the intersection of the bearing lines from each of them will locate the target.

A major problem of passive sonar is that of noise from the detecting vessel and from other vessels that are not the target. One way of reducing this problem is to tow the passive sonar equipment at some distance behind the operating vessel. Another source of difficulty is the presence of a thermocline, a layer in the ocean where the temperature gradient is higher than in the water both above and below. When sound waves strike a thermocline layer they are partially reflected and scattered, which provide extraneous sources of noise that hinder the process of locating and identifying the target.

Submarines tend to employ passive rather than active sonar since they are not then emitting noise, i.e. pings, that can be used to track them. If they are in silent-running mode, either powered by a nuclear reactor or batteries, then they will avoid the use of pumps or any other machinery that inevitably will emit a characteristic noise. Another noise source that could give away the presence and location of submarines is that of their propellers as they beat against the water, sometimes causing cavitation — the formation of bubbles that generates a characteristic sound. By careful design the noise from propellers can be greatly reduced.

Passive sonar has other forms of military application. Sea mines can be fitted with passive sonar detectors rather than the more customary explosive detonators that look like spiky protuberances on the surface of the mine. The sonar detects the noise of the passing ship and can detonate when the ship is directly overhead. Since they can be well below the surface they are more difficult to detect. They can also be difficult for minesweepers to deal with as they can be made to explode only when a second ship passes overhead so a single sweep will do no more than activate the mine for the next ship that passes over it. The major powers have also established passive sonar buoys in various ocean locations that can detect and identify ships in their vicinity, thus enabling the operating countries to keep tabs on the dispositions of naval vessels of other nations.

13.5. Imaging the Interior of the Earth

Knowledge of the internal structure of the Earth has to be gained by indirect means. The Earth has a mean radius of 6,371 km and the deepest mine created by man, the Tau Tona gold mine in South Africa, penetrates to a depth of 3.9 km — just a scratch on the surface. Our knowledge of the Earth's interior comes about from a study of waves, known as *seismic waves*, which pass through the Earth and are caused by earthquakes. Where this phenomenon occurs in populated areas of the Earth it can cause severe loss of life and structural damage, either directly by violent movements of the Earth or through the effect of tsunamis — huge tidal waves — as they sweep over low-lying areas. However, out of the tragedy of earthquakes there is some positive outcome in the form of knowledge about the structure of the Earth.

13.5.1. *Types of seismic wave*

There are two main classes of seismic wave: those that only travel along the Earth's surface and those that pass through the body of the Earth. The two kinds of surface wave, *Love Waves* and *Rayleigh Waves*, differ in the way that near-surface matter vibrates but both give a surface motion rather like that of waves moving over the sea. These are the types of wave that are most destructive, the ones whose effects we see but, since they do not travel into the deep interior of the Earth, they do not concern us here. We are interested in the two types of wave that *do* travel through the Earth — known as *body waves*. The vibrations from a major earthquake occurring anywhere can be detected by seismic stations situated all over the world in widely separated locations.

The two types of body wave — *P-waves* and *S-waves* are illustrated in Fig. 13.3. P-waves (primary waves) are longitudinal waves, i.e. the vibrations of particles of matter constituting the wave, like the sound waves described in §13.1 and illustrated in Fig. 13.1, are vibrating to and fro in the direction of propagation. These waves can travel through solids or liquids, both of which react elastically to compression by re-expansion. In Fig. 13.3 what is shown for a P-wave is the passage of compressions and dilatations (rarefactions) in the direction of propagation.

The motion of material in S-waves (secondary waves) is transverse, i.e. perpendicular to the direction of propagation. The material is experiencing shear and liquids have no resistance to shear — if neighbouring layers of water slide past one another then, unlike solids, they experience no restoring force taking them back again. For this reason S-waves can only pass through solid material. They move

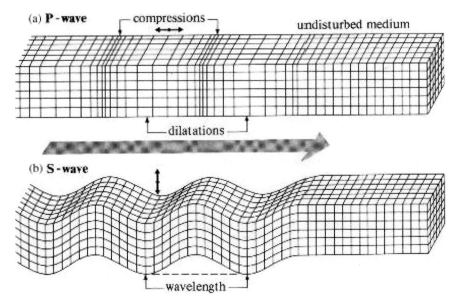

(a) **P-wave**   — compressions —   undisturbed medium

—dilatations—

(b) **S-wave**

—wavelength—

Figure 13.3   Body waves.

at a slower speed than P-waves, typically about $3\,\mathrm{km\,s^{-1}}$ in granite compared with $5\,\mathrm{km\,s^{-1}}$ for P-waves in the same material.

13.5.2. *The passage of body waves through the Earth*

All wave motions of any type, electromagnetic or mechanical, have the characteristic that their passage is modified by variations of their velocity in the material in which they are moving. Within the Earth the variation of velocity can be of two forms. Because the pressure on terrestrial material, and hence its density, varies with depth so does the velocity of a wave motion travelling within it; in general velocity increases with density. The effect of this is that a wave travelling through a medium, which may be chemically homogeneous, will have its direction of motion constantly changing by what we can think of as continuous refraction. This is illustrated in Fig. 13.4(a). The other form of variation is when a wave hits a sharp discontinuity. Where the wave is moving from where the speed is less to where the speed

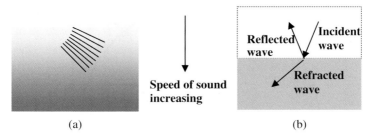

(a)   (b)

Figure 13.4   Behaviour of a seismic wave (a) with speed continuously varying and (b) with a sharp discontinuity of speed.

is more, then there are two effects. The wave is diffracted to travel in a direction further from the normal to the interface and also some part of the wave is reflected backwards into the medium whence it came. This is shown in Fig. 13.4(b). Conversely, if the travel is from a region of greater speed to one of lesser speed then reflection will still occur but the refraction will deflect the wave closer to the normal.

Although the inner structure of the Earth has been discovered by the use of seismic waves, for the purpose of explaining how seismic wave observations are interpreted it is better first to describe the structure of the Earth and then to show how the waves travel through it as a result of the structure. The general structure of the Earth is shown in Fig. 13.5. Starting from the surface the first layer is the *crust*, on average 35 km thick and consisting of low-density rocks. There are large variations of both thickness and density. Oceanic crust is between 6 and 11 km thick and consists of basaltic rock that continually flows up through cracks in the crust to form new sea floor; its average density is about 3,000 kg m^{-3}. Continental crust is between 30 and 75 km thick and consists of igneous rock of lower density, about 2,700 kg m^{-3}. It consists of the lighter rocks floating on the surface when the whole Earth was molten and the oldest continental rocks, situated in Greenland, are 3.8 billion years old. A schematic illustration of the structure and variation of the crust is shown in Fig. 13.6.

The mantle is a shell of denser rocks accounting for 84% of the Earth's volume and about two-thirds of its mass. The chemical

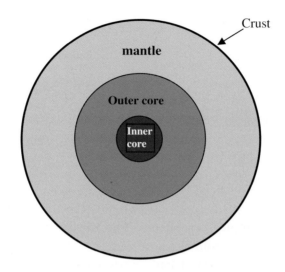

Figure 13.5   The internal structure of the Earth.

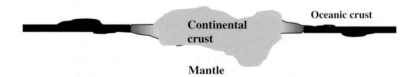

Figure 13.6   The structure of the Earth's crust.

composition is not uniform since some minerals are unstable either above or below certain pressures and, since the pressure, density and temperature increase steadily with depth, the speed of sound also varies, in general increasing with depth. The crudest division of the mantle is into *upper mantle*, extended to a depth of about 660 km and the remainder called the *lower mantle*. In the upper mantle the high temperature makes the rocks slightly plastic, which means that over long periods they can flow and set up convection currents. It is motions due to these currents that exercise drag forces on the crust above and produce *continental drift*, whereby large coherent crustal plates move relative to one another — the source of the earth-quakes that both cause devastation and enable us to explore the inner Earth. Although the lower mantle is at an even higher temperature, because of the very high pressure, and perhaps because of chemical differences, it is less ductile than the material immediately above it.

The core consists predominantly of iron with substantial amounts of nickel. Some of the iron may be in the form of the mineral *troilite*, iron sulphide, FeS. In the outer core, with radii between 1,210 km and 3,470 km the combination of temperature and pressure is such that it consists of molten material. It is in this shell, mainly of molten iron, where convection currents produce the dynamo effect that gives the electric currents that generates the Earth's magnetic field. Although the inner core is at an even higher temperature the pressure is so high that it is solid.

With this background we can now understand how body waves travel within the Earth. First, in Fig. 13.7, we look at some ways in which body waves can travel via the mantle from the focus (earthquake source), shown as F, some distance below the surface, to the seismic station, situated at Q. The simplest path is the direct one, labelled 1, and the first disturbance at Q will be the P-wave following this route. Notice the curvature of the path due to continuous refraction. Other routes are possible: the route labelled 2 is in two parts and involves reflection at the interface between the mantle and crust, with some waves penetrating through the crust and being reflected at the crust-atmosphere interface. The P-wave travelling via this route will

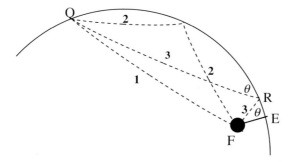

Figure 13.7   Some paths from the focus of an earthquake to a seismic station.

arrive later than the one from route 1 but probably earlier than the slower-moving S-wave from route 1. Route 3, another in two parts, is reflected from region R, close to E, the earthquake epicentre, which is the point on the surface directly over F. The difference on the times of arrival of the P- and S-waves via R enables an estimate to be made of the depth of the earthquake, an important parameter in determining its destructive power.

Figure 13.7 shows the general form of the different paths from F to Q that applies to both P- and S-waves although, because of the different speeds of the two kinds of wave, their actual paths will differ. For waves penetrating deeper, ones that reach the liquid outer core, the behaviour of the two waves is quite different and depends on the way that their speed varies with depth, shown in Fig. 13.8.

In Fig. 13.9(a) the paths of various P-waves are shown. When they strike the boundary between the mantle and outer-core they are refracted and reflected; only the refracted wave that passes through the boundary is shown in the figure. The path marked 1 is a critical path that is just tangential to the liquid core surface but stays within the mantle. It eventually reaches the surface at point A. A slightly

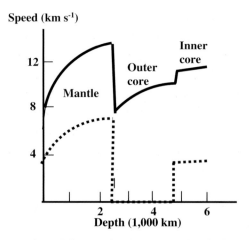

Figure 13.8   The speed at different depths in the Earth for P-waves (full line) and S-waves (dotted line).

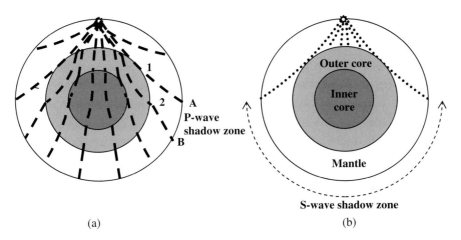

(a)   (b)

Figure 13.9   Possible paths through the Earth of (a) P-waves and (b) S-waves.

displaced path that strikes the liquid core surface obliquely will be refracted and then further refracted at other boundaries as it travels, until it emerges at the surface of the Earth at B some distance from A. If path 2 is the critical path that just gives transmission at the mantle–inner-core interface then no P-wave will reach any seismic station situated anywhere in the region between A and B. This region is known as the *P-wave shadow zone*: it covers all points of the Earth for which radii to the centre of the Earth from epicentre and from the point subtend an angle between 105° and 140°.

For S-waves we have to take account of the fact that they do not travel through fluids and hence cannot penetrate the inner core region. Any waves travelling through the mantle that strike the inner core surface are absorbed and travel no further. The possible paths of S-waves are shown in Fig. 13.9(b). In this case the *S-wave shadow zone* covers about one-half of the Earth, anywhere with angular distance from the epicentre, as defined for the P-wave shadow zone, more than about 90°.

This description of the paths of P- and S-waves is somewhat simplified but gives the flavour of the main pattern. In practice, waves can be reflected to and fro between boundaries to give rather complicated disturbances at the seismic stations.

13.5.3. *Interpretation of seismic wave data*

The instrument that detects earthquake waves is the *seismometer* that, when it gives a plotted output, is called a *seismograph*. In its simplest form a seismometer is a frame, rigidly attached to the Earth so that it follows its local motion, from which is suspended an object of such large mass that its inertia prevents it from following the motion of the frame. The relative motion of the frame and the massive object gives a measure of the Earth's motion. Such an instrument can measure with great accuracy vibrations due to P- and S-waves and surface waves, where the frequency of the waves is of the order of seconds. There is another type of seismic disturbance we have not mentioned, *full-body waves*, that involve the vibration of the whole Earth and have frequencies of the order of hours. Three kinds of full-body waves are illustrated in Fig. 13.10. In Fig. 13.10(a) the motion is one where two hemispheres execute twisting vibrations in the opposite directions to each other. Figures 13.10(b) and (c) show two possible kinds of long wavelength waves travelling round the Earth.

Seismic stations measure the arrival of the different P- and S-waves representing the total wave package; P-waves arrive first, followed by the slower S-waves. The difference of time of arrival of the first P- and S-waves, which travel by the direct route from focus

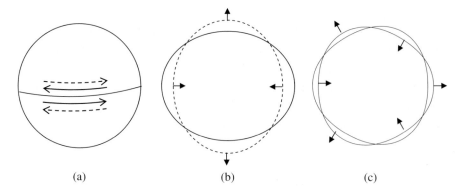

(a)   (b)   (c)

Figure 13.10   Three kinds of full-body wave.

to seismic station, is a measure of the approximate distance of the earthquake — the estimate depending on its unknown depth — but not its direction. However, if three or more stations record the earthquake then by triangulation its location can be determined. This assumes knowledge of the speeds of seismic waves at different levels of the Earth but before this was known the problem that Earth scientists had to solve was how to find the speeds at different levels from their seismic data.

A general approach to finding a model for the Earth's interior is to assume a model for the speed of sound at different depths and then to infer the form of wave travel and the expected wave patterns at various seismic stations. From the difference in recorded and inferred patterns the model can be refined. A factor that has to be taken into account is the dissipation of sound waves as they travel through the Earth. Dissipation increases with the degree of liquidity or plasticity of the material that, for a particular material, depends on its temperature and pressure. The first work on interpreting seismic data in terms of Earth structure was very early in the twentieth century by the German geophysicists E.J. Wiechert (1861–1928) and K. Zoeppritz (1881–1908), who restricted their studies mostly to disturbance from shallow and close earthquakes. This approach was greatly extended in work begun by the British geophysicists K.E. Bullen (1906–1976) and H. Jeffreys (1891–1989), who created the first models of the Earth's interior from seismic data, based on estimates of the speed of sound at different depths. With the advent of computers that enable large numbers of Earth models to be quickly tested against the seismic data, knowledge of the Earth's interior has been greatly extended and refined.

The Apollo astronauts who landed on the Moon between 1969 and 1972 left seismometers that transmit their readings to Earth. There are numbers of low-powered moonquakes, and occasional meteorite strikes on the Moon also create seismic disturbances. From the information received it has been possible to produce a model of the Moon's interior, which shows that it has an iron-nickel core just less than 400 km in radius.

13.5.4. *Geoprospecting with sound*

Large energy events, such as earthquakes or, in the past, tests of nuclear weapons, enable the internal structure of the whole Earth to be explored. The natural resources that are exploited by mankind, e.g. oil and iron ore, occur underground, sometimes at depths of hundreds of metres to kilometres, and give no indication of their presence on the surface. While it is suspected that a general area may be a likely source of the desired mineral, to detect the probable location without the expense of random exploratory drilling is obviously desirable.

Producing an explosion with a modest amount of high explosive, such as dynamite used extensively in mining, can produce local seismic waves that penetrate sufficiently far into the crust to be reflected and refracted by an interface between materials with different sound speeds. By monitoring the waves received in various locations around the explosion site it is possible to detect the presence of boundaries between different kinds of material in the upper crust close to the source of the explosion. For example, oil and natural gas are often found in pockets within salt domes, large bodies of salt left behind by the evaporation of extinct seas. The speed of sound in salt is in the range 4–6 km s^{-1}, normally different from that of the surrounding rocks. Sound waves will be reflected whenever they strike an interface between salt and rock so enabling the profile of the salt dome to be determined. There are other physical techniques for geoprospecting and the use of multiple techniques, when they can be applied, can give a better picture of underground conditions than any single one on its own.

Chapter 14

Medical Imaging

14.1. The Discovery of X-rays

At the end of the nineteenth century, experiments involving electrical discharges through vacuum tubes, which contained some residual gas, were very much the vogue, especially with German and English scientists. Something passed from the cathode (negative electrode) to the anode (positive electrode) and caused the glass beyond the anode to fluoresce. This unknown entity was given the name *cathode rays* and it was generally believed, especially by German scientists, that they were some form of radiation but, since it could be deflected by a magnetic field, it would have to be radiation of a previously unknown kind.

The other possibility was that cathode rays are particles with a negative charge, emitted by the cathode and then attracted towards the positively charged anode. William Crookes (1832–1919), an English chemist, suggested that these particles were gas molecules that had picked up a negative charge from the cathode. He designed a new kind of tube, the *Crookes tube* (Fig. 14.1), to show the fluorescent glow. The cross-shaped anode is opaque to cathode rays and its shadow image is visible on the phosphor-coated anode. By placing a magnet in the vicinity of the tube the deflection of the image on the screen can be demonstrated.

Later, in 1897, the English physicist, J.J. Thomson (1856–1940) showed that cathode rays consist of the light negatively charged particles that we now know as electrons; for this discovery Thomson was awarded the Nobel Prize for Physics in 1906.

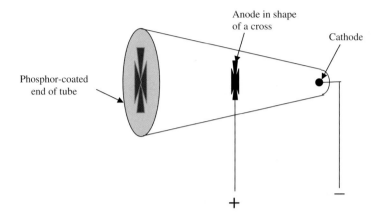

Figure 14.1   A schematic Crookes tube.

Figure 14.2   Wilhelm Conrad Röntgen.

One of the individuals investigating cathode rays prior to Thomson's discovery was the German scientist, Wilhelm Conrad Röntgen (1845–1923; Fig. 14.2) — by all accounts a quiet and modest man. Röntgen was using a vacuum tube incorporating an aluminium-foil window behind the anode, thin enough to allow the cathode rays to escape. Röntgen reinforced the flimsy foil with cardboard that was also thin enough to transmit the cathode rays. When another piece of cardboard, coated with barium platinocyanide, a fluorescent material, was placed near the end of the tube it glowed. Vacuum tubes emit some visible light because of excitation of the gas molecules they contain — the basis of the displays now used for shop signs and advertising, e.g. neon signs. The cardboard reinforcement for the foil should have prevented any light from coming from the tube to produce the fluorescence so it looked as though it had been produced by the cathode rays. As a good scientist Röntgen decided to make sure of his conclusion so he covered the whole tube in black cardboard to make it light tight and looked for any trace of light with the tube running in a darkened room. No light came from the tube but there was a faint glow from a source some distance from the tube — too far for the cathode rays to travel. Röntgen found that the glow was due to fluorescence of the barium-platinocyanide-covered cardboard strip.

Röntgen concluded that some new form of radiation was responsible and he investigated its properties. He found that these rays, which he called *x-rays* to indicate their uncertain nature, blackened a photographic plate. Experiments that measured the extent of blackening of a photographic plate showed that the transmission of x-rays depended on the type and thickness of the material through which they passed. Because bone and flesh absorb differently he found that the bones of a hand could be imaged photographically by passing x-rays through it. A radiograph of a colleague's hand is shown in Fig. 14.3. Medical radiography quickly became, and still is, an essential diagnostic technique. For the important discovery of x-rays Röntgen was awarded the first Nobel Prize for Physics in 1901.

It took some time for the danger of overexposure to x-radiation to be discovered and x-ray generators were sold freely as a means of home entertainment to look at bones within the body; there must have been many premature deaths due to this cause.

14.2. X-ray Generators

X-rays are a form of electromagnetic energy of extremely short wavelength, typically 0.1 nm.[1] In many applications of x-rays it has

---

[1] 1 nm (nanometre) is 10^{-9} m.

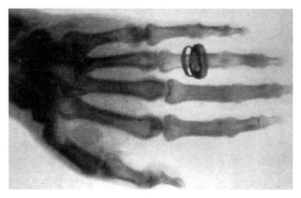

Figure 14.3   A radiograph taken by Röntgen of the hand of Albert von Kolliker.

been found convenient to express wavelengths in terms of another unit, the Ångstrom unit (Å) that is $0.1\,\text{nm} = 10^{-10}\,\text{m}$. The simplest way to produce x-rays is to fire a beam of very high-energy electrons at a metal target, one preferably made of a heavy metal. The electrons, colliding with the metal atoms, lose energy in one or more interactions and each energy loss is converted into a photon, a packet of electromagnetic energy. The energy of a photon of wavelength λ is given by

$$E = h\nu = \frac{hc}{\lambda}, \tag{14.1}$$

where h is *Planck's constant*, $6.626 \times 10^{-34}\,\text{Js}$, ν is the frequency and c is the speed of light, $2.998 \times 10^{8}\,\text{ms}^{-1}$. From this equation the energy of an x-ray photon of wavelength $1\,\text{Å}$ is found to be $1.99 \times 10^{-15}\,\text{J}$. It is customary in many scientific contexts to express atomic-scale energies in units of *electron volts* (eV), the energy of an electron that had been accelerated by a potential difference of 1 volt. This unit is equivalent to $1.602 \times 10^{-19}\,\text{J}$, giving the energy of a $1\,\text{Å}$ photon as $12.42\,\text{keV}$.[2] For x-rays to contain a component with wavelength as short as $1\,\text{Å}$ the electrons must be accelerated through a potential difference of at least $12.42\,\text{kV}$ and for shorter wavelengths even higher potential differences are required.

---

[2] $1\,\text{keV} = 1{,}000\,\text{eV}$.

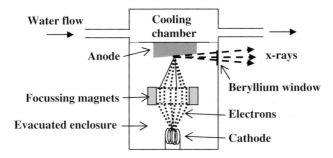

Figure 14.4   A schematic medical x-ray generator.

The basic design of an x-ray generator is shown in Fig. 14.4. A hot cathode provides the source of electrons that are accelerated towards the anode that, for medical applications, is often made of tungsten, the heavy atoms of which are very efficient in slowing down electrons. A potential difference of order $100\,\mathrm{kV}$, produced by a high-voltage generator, is maintained between the cathode and anode. To give a compact source of x-rays the beam of electrons is focused onto a small spot on the anode. With a current of a few milliamperes (mA) passing through the generator the total power being consumed is the product of current and potential difference and is of order $1\,\mathrm{kW}$. Much of this appears as heat generated at the small focal spot on the anode so, to prevent damage to the anode, it is cooled by a flow of water. The flat surface of the anode is tilted so that x-rays from the focal spot can pass through a beryllium window to the outside of the generator. Beryllium is a metal of very low density and a thickness that can withstand the external pressure on the evacuated generator will also allow the free passage of the emerging x-ray beam.

X-ray generators are variable in their design and construction. Rarely will they externally resemble what is shown in Fig. 14.4, which is just a schematic representation of the essential structure of a generator. Within a hospital environment they will be incorporated into large pieces of equipment that are designed for specific kinds of diagnostic technique, where the generator itself can be moved to image different parts of the patient's body, usually in a prone position. At the other end of the size scale, although not designed for medical use, there are battery-powered small mobile generators that can be carried easily and are used for testing the integrity of large structures, for example, bridges or mobile cranes, by looking for cracks within metal components of the structure.

14.3. Recording a Radiographic Image

The danger of overexposure to x-radiation has already been mentioned. Very high-energy photons can disrupt DNA in cells and produce harmful mutations that cause various kinds of cancer. In normal life there is background radiation from various sources so small doses of x-radiation for diagnostic purposes do not appreciably add to the overall risk but large doses are best avoided. For this reason most x-ray images are produced with short bursts of radiation lasting only a few microseconds.

The traditional way of recording a diagnostic x-ray image is with film. The desirable film characteristics are that it should be high-speed, so that fewer photons are required to produce the image thus reducing the patient's exposure, and also that it should give a resolution that matches the diagnostic requirements. Unfortunately these properties are usually incompatible — high-speed films also tend to give lower resolution. For this reason it is normal to use x-ray film in contact with a screen coated with a phosphor that converts the impinging x-radiation into visible light to which the film is more sensitive.

Another way of viewing a weak image is with an x-ray image intensifier, the form of which is illustrated in Fig. 14.5. A low-intensity x-ray image is formed on an input screen that has the property that it is conducting, so that it can serve as a cathode, and also has its inner surface coated with a material such as caesium iodide (CsI) that readily emits photoelectrons — electrons ejected from the CsI by x-ray photons. The energy of the photoelectrons is greatly increased by accelerating them through a large potential difference, typically $30\,kV$, while at the same time focusing them to all move through the same point so that the spatial distribution of electrons retains the form of the image. The high-energy electrons then

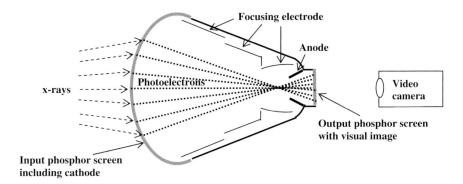

Figure 14.5   An x-ray image intensifier.

impact the output screen coated with a phosphor that absorbs their energy and converts it into visible photons. This produces a bright visual replica of the original weak x-ray image, which can be directly viewed, photographed or stored by the use of a CCD video camera.

The essential requirements of x-ray medical imaging are to acquire diagnostic information while at the same time keeping the x-ray dose to a minimum. For simple single-shot x-ray exposures this means choosing x-rays of energy appropriate to the application — for example, for dental x-rays where the interior of the teeth are being examined, hard, i.e., high-energy, short-wavelength x-rays are required. For many procedures, such as installing a pacemaker where wires are inserted via a vein into the heart, it is necessary constantly to monitor the passage of the wires, which is best done with x-rays. However, this requires an extended period of exposure, the intensity of which can be reduced by the use of an image intensifier so that the total x-ray dose is kept within acceptable bounds.

Another development in medical x-ray imaging uses array detectors consisting of a matrix of amorphous silicon detectors. The incoming x-radiation is converted into electrical impulses by one of the two techniques illustrated in Fig. 14.6. In Fig. 14.6(a) the x-ray photons fall on a layer of scintillating material and generate pulses of visible light when they are absorbed. This light is received by the

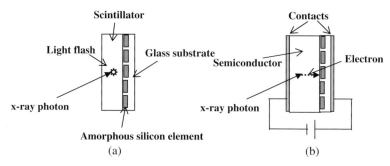

Figure 14.6   X-ray imaging using amorphous silicon arrays with (a) a scintillator and (b) a semiconductor.

tiny amorphous silicon element that converts it into an electrical signal that is sent to a computer and stored. The problem with this system is that the light pulse can influence more than one element so that the resolution is somewhat reduced. Having a very thin layer of scintillator can minimize the reduction of resolution, but this is at the expense of reducing the conversion efficiency of x-rays to light. In practice a compromise is reached between the requirements of efficiency and resolution.

The resolution problem can be resolved by replacing the scintillator with a layer of a heavy-atom semiconductor, coated with a conductor on the input surface (Fig. 14.6(b)). X-rays are absorbed by the semiconductor and generate electrons that are accelerated directly towards the silicon elements by an applied potential difference. As for the previous method, the generated electrical signal is passed to a computer.

14.4. Computed Tomography — CT Scans

The word *tomography* is derived from two Greek words that together mean 'writing a slice' and, in medicine, is concerned with producing a representation of the structure within a plane section of a body. An Italian radiologist, Alessandro Vallebona (1899–1987), first proposed a way to do this in the 1920s and it was a technique widely used until about 1970, when it was largely replaced by a superior procedure, *computed tomography*.

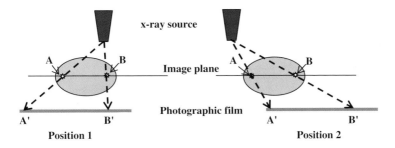

Figure 14.7   The principle of tomography.

The tomography process is based on simple geometrical ideas and is illustrated in Fig. 14.7. The x-ray source gives a divergent beam that irradiates the whole of the body and falls on the photographic film giving a projected image of the x-ray absorption of the body. In position 1 the rays passing through points A and B in the image plane of the body fall, respectively, on the points A′ and B′ of the film. In position 2 the x-ray source and film have moved in such a way that rays passing through points A and B still pass through A′ and B′. In moving between positions 1 and 2 this relationship is always preserved and, from geometrical considerations, it will always be true for all points in the image plane. For any other plane in the body this relationship is not true so the total effect of the complete motion is that there is a sharp image for the image plane and blurred images for all other planes, which just contribute a background noise. While the quality of tomography images were not high by present standards, they were good enough to provide useful diagnostic information.

The theory behind computed tomography is quite different — highly mathematical and involving a concept known as the *Fourier transform*[3] (FT) — but here it will only be described in general descriptive terms. Many physical entities can be described by a distribution of some quantity, which we take here as a density. In the case of medical x-ray diagnosis, the density of interest is that of the bodily tissue for absorbing x-rays. For any such function there

---

[3] A full discussion of the Fourier transform can be found in M.M. Woolfson (1997) *An Introduction to X-ray Crystallography*, 2nd edn, Cambridge, Cambridge University Press, pp. 76–105.

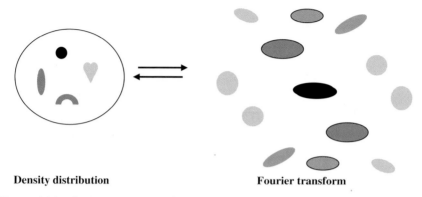

Density distribution   **Fourier transform**

Figure 14.8   A representation of a two-dimensional density distribution and its Fourier transform.

can be defined an FT, a function with the same number of dimensions as the density distribution, which has the property that a complete knowledge of the density distribution fully defines the Fourier transform and *vice versa*. The value of the FT at each point is, in general, a complex quantity with an amplitude and phase like the individual aerial contributions shown in Fig. 7.12. The relationship between a density distribution and its FT is schematically illustrated in Fig. 14.8 for a two-dimensional density distribution. Each point of the FT has a magnitude (illustrated by variable shades of grey) and a phase that can be anywhere in the range 0 to 2π. The arrows show that, by mathematical means, each can be derived from the other.

In Fig. 14.9 we now show all the density projected onto a line. Since the projected density is one-dimensional then so is its FT and the essential basis of computed tomography is that its FT is just a *line of the two-dimensional FT* — that through its centre and parallel to the line on which the density is projected.

The process of computed tomography, illustrated in Fig. 14.10, depends on this relationship between a projected density and its FT and involves the following steps:

(1) With a fine x-ray beam scan along a line in the plane of the required section, measuring the intensity of the transmitted beam with a moving detector to measure the transmitted beam.

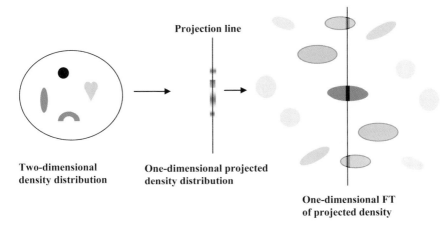

Figure 14.9   The one-dimensional FT of density projected onto a line.

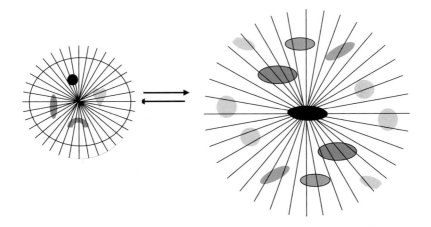

Figure 14.10   Projection lines in a two-dimensional density distribution and corresponding sampling lines of the two-dimensional transform.

This gives the projected absorption density on a line in the section.

(2) Compute the one-dimensional FT of the density. This is a sample of the two-dimensional transform along a line. This FT is stored.

(3) Repeat this process but each time move the projection line through a small angle, usually 1° (for clarity 5° in the figure).

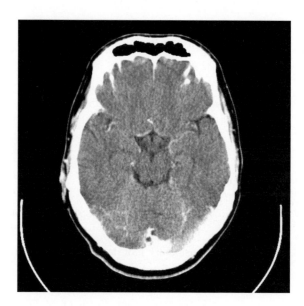

Figure 14.11   A section of the brain obtained from computed tomography.

(4) Combine the one-dimensional FTs to give a good representative sample of the continuous two-dimensional FT.

(5) Compute the two-dimensional density in the section from the two-dimensional FT and output as a density distribution on a CRT or similar display unit.

This procedure gives images of sections of the human body of high quality without the noise accompanying simple tomography. Figure 14.11 shows a section through a skull with details of the brain's structure clearly revealed. A series of sections, starting at the base of the skull and moving upwards, gives a complete picture of the whole structure of the brain that would clearly indicate any anomalies.

In 1968 — somewhat before the medical applications of CT became established — the British scientist, Aaron Klug (b. 1926; Nobel Prize for Chemistry, 1982), made an interesting, but different, application of the technique of image reconstruction using FTs. Electron micrographs give a projected image through small specimens such as bacteria. Since the appearance of the image depends on its orientation, it is difficult to determine the three-dimensional structure of a bacterium just by inspection of randomly oriented projections. For a particular orientation Klug found the two-dimensional FT of a projection of the density of a bacterium and then by tilting the specimen holder was able to find the two-dimensional FTs for a whole succession of orientations. Together these gave a reasonable sample of the three-dimensional FT corresponding to the density structure of a bacterium. Then, by transforming the three-dimensional FT into the equivalent density, he was able to produce a three-dimensional model of the bacterium. Actually, because of radiation damage to the specimen when it was repeatedly exposed, he used several different bacteria to sample the whole of the three-dimensional FT, but the principle is as described.

14.5. Magnetic Resonance Imaging

Magnetic resonance imaging (MRI) gives the same kind of information as CT but with the advantage that no potentially damaging radiation is involved, so reducing risk to the patient. However, from our point of view there is an accompanying disadvantage — that the theory of how it works involves some fairly difficult physics, all happening at the atomic level for which there is no everyday experience to guide us. The following account is, perforce, somewhat simplified but suffices to explain the basic principles.

The working of MRI depends on a property of the proton, the nucleus of the element hydrogen that is contained in water, a major component of the body, and in many other body materials. The property that is relevant here is that it behaves like a tiny magnet, with a north pole and south pole just like the needle magnet in a compass. Under normal circumstances, with no external magnetic field present, the magnets orient themselves in random directions. In the case of a compass-needle magnet we know that it always lines up with the Earth's magnetic field and indicates the direction of the Earth's magnetic north pole but atomic magnets, like those of protons, when placed in a steady magnetic field behave rather differently. Some

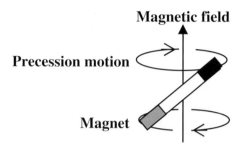

Figure 14.12   Precession of a magnet around a magnetic field direction.

do line themselves parallel with the field, as we would instinctively expect from the behaviour of a compass needle, but, rather paradoxically, some of them align themselves in the opposite anti-parallel direction. There is an excess of magnets parallel to the field, the extent of the excess increasing linearly with the strength of the field. The excess is small; for a field of 1 T (30,000 times that of the Earth)[4] for every million protons in the parallel state there are 999,994 that are anti-parallel. However, it is the small excess in the parallel direction that enables us to obtain an MRI signal. An important characteristic of these two possible alignments is that the anti-parallel one contains more energy; to move a magnet against its natural inclination from the parallel to the anti-parallel state requires an input of energy.

It is a characteristic of a magnet that if it is inclined to the direction of an externally applied magnetic field it undergoes *precession*, a rotation of the magnet axis around the direction of the field (Fig. 14.12). The frequency of the rotation does not depend on the angle of tilt between the magnet axis and the field direction and in the case of the proton, or for atomic nuclei in general, this frequency, ω, is known as the *Larmor frequency*, given by

$$\omega = \gamma B, \tag{14.2}$$

---

[4]T is the symbol for *tesla*, the SI unit of magnetic field. It is equivalent to 10^4 gauss, the cgs unit.

where B is the strength of the magnetic field and γ is the *gyromagnetic ratio*, a characteristic of the type of atomic nucleus, in our case of interest a proton.

For a magnetic field of 1 T the precession frequency for a proton is 42.56 MHz. If a radio-frequency electromagnetic field with *exactly* the Larmor frequency is applied to the protons then some parallel protons absorb energy. They tilt away from the parallel configuration and begin to move over towards the higher-energy anti-parallel direction. The extent of this tilting process depends on both the intensity of the radio wave and its duration.

The next stage is to switch off the radio-frequency field so that the input of energy is removed. Now the system of magnets is not in equilibrium and the tilted protons begin to return to their original parallel configuration to give the original equilibrium proportion of parallel and anti-parallel protons. As they do so they release electro-magnetic radiation at the Larmor frequency. Some of this energy is absorbed by, and heats, the tissue within which it is produced, but most escapes to the outside and provides the signal for the MRI scan. If the maximum tilt of the proton magnets is less than 90° then the variation of the intensity of the output signal with time is of a declin-ing exponential form, as shown in Fig. 14.13. There are actually two separate processes contributing to this decline — one we may think of as just the natural inclination of the proton magnets to return to an equilibrium configuration and another, called *spin-spin interactions*,

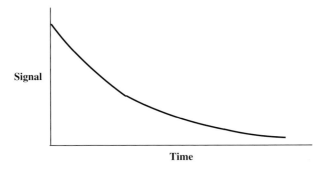

Figure 14.13 A declining signal with time after the radio-frequency excitation is removed.

Table 14.1   T1 and T2 values for some body tissues.

Material	T1 (ms)	T2 (ms)
Muscle	870	47
Kidney	650	58
Grey matter	920	100
Lung	830	80
Liver	490	43

is where neighbouring protons exchange energy so that one goes from parallel to anti-parallel while the other does the reverse. This leads to different delays in the electromagnetic outputs of different protons so that they are no longer in phase, giving destructive interference. This effect increases with time so giving decay of the output signal. Both the decay processes have timescales dependent on the type of material involved, designated by times T1 and T2 respectively, which are the times for the output to drop to $1/e$ of the original value. Some typical T1 and T2 values are given in Table 14.1.

Because there are two separate decay times involved, if we compared the returning signals from two regions with different materials then the relative strengths of the total signals will vary as a function of time from when emission begins. The measurement is of the cumulative signal from the beginning of the emission stage and the relative values from two regions will depend both on the density of hydrogen in the material and the time the signal is collected. Typically, for good tissue discrimination a time of about 0.4 seconds is used. These accumulated signals, displayed on a CRT with intensity proportional to the total signal, show the difference in the two regions. However, we must now consider how spatial discrimination occurs in an MRI scan, which requires a description of the construction and operation of an MRI machine. An MRI scanner is shown in schematic form in Fig. 14.14. The patient is inserted on a sliding platform into a large cylinder. Not shown in the figure, since they are situated within the walls of the cylinder, are a number of coils, electric currents through

Figure 14.14　A schematic MRI scanner.

which set up magnetic fields of various kinds within the cylinder. Firstly there are the *main coils* that give a large (0.2–3 T) uniform field throughout the length of the tube. For high fields it may be necessary to use superconductor magnets where the coils are cooled to liquid helium temperature so that their electrical resistance falls to zero and, hence, large currents can be passed through them without excessive heating. Then there are three sets of *gradient coils* giving much smaller fields with a linear variation in the x, y and z directions. The linear gradients are usually of the order $0.2\,\mathrm{mT\,m^{-1}}$.

Now we consider how we would perform an MRI scan in the xy plane at a particular cross-section of the patient. First we switch on the main coils and also the z-direction gradient coils so that there is a high, but linearly varying field in the z direction. Since B varies linearly along the z direction then, from Eq. (14.2), the Larmor frequency also varies in a linear fashion. Now we excite the proton magnets with a pulse of electromagnetic radiation of frequency that corresponds to the Larmor frequency at the z position where the scan is to take place. The only proton magnets to be excited are those at, or very close to, the cross-section of interest and the thickness affected can be reduced or increased by, respectively, having a higher or lower field gradient. When the exciting pulse is switched off the z gradient is removed and the x-gradient coils are immediately switched on. Now the Larmor frequency varies in the x-direction and the frequency of the radiation being released by the proton magnets as they return towards their equilibrium state is just x-dependent. What the radio-frequency detector picks up, in the 0.4 seconds or so it is collecting the signal, is a range of frequencies, the intensity at each frequency being the combined contribution of the signal from the hydrogen content of the complete range of values of y at a particular value of x. This is like the data from a single CT scan and its one dimensional FT is a sample of the two-dimensional output of the complete $x-y$ cross-section. The whole process is now repeated but with the modification that the x-gradient field is slightly reduced and a small y-gradient field is added so that the net field gradient in the $x-y$ plane is inclined at $1°$ to the x-axis. The FT is now a sample of the complete two-dimensional FT on a line making an angle of $1°$ with the original one. By repeating this process, each time changing both the x-gradient field and the y-gradient field to rotate the field gradient until the x-gradient field is the reverse of its initial value, a sample of the complete two-dimensional FT is found. This is transformed to give an image of the two-dimensional section of the patient — which is the MRI image.

The procedure described above is not the best or quickest but it has the virtue of being the simplest to understand and it fully describes the basic physical principles involved. The flexibility of having three independent intensity gradients is that it is possible to obtain scans in cross-sections at any angle to the axes. By varying the collecting time of the signal it is possible to emphasize either the T1 or T2 components from a particular type of tissue and it is clear from Table 14.1 that this can give different contrast between pairs of tissues. Figure 14.15 gives T1 and T2 brain scans that show the differences of contrast that can be obtained.

For the most part CT has been replaced by MRI, which gives better results without the complication of radiation exposure. It has some disadvantages; patients with claustrophobia find it difficult to remain within the machine for the 15 minutes or so taken to complete the scanning process and while the scanning is taking place the magnets can be very noisy — it has been described as like being in a metal dustbin while people outside are beating it with hammers. Another problem is that one must ensure that no metal objects are present within the machine while it is running since they can be projected at high velocity when the large magnetic field is switched on and cause serious injury or even be fatal. Finally, it is unsuitable for use on any patient with a magnetic-metal implant or with a

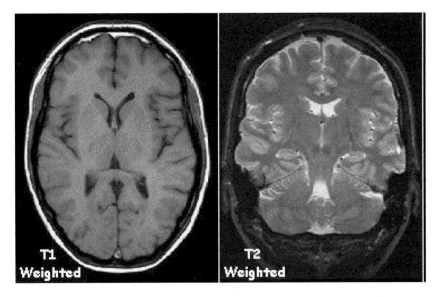

Figure 14.15   MRI brain scans taken with T1 and T2 signals.

pacemaker, the working of which would be affected by the magnetic field.

14.6. Imaging with Ultrasound

Chapter 13 described various aspects of the use of sound for detection and imaging but missed one important area — medical imaging with sound. This branch of medical science had its origins in America in the late 1940s with the work of a British doctor and medical scientist John Wild (1914–2009), who developed equipment first to measure the thickness of the bowel wall in 1949 and then, a few years later, to detect tumours in the breast and the brain.

The outstanding characteristic of the sound, called *ultrasound*, used for medical imaging is its very high frequency, mostly in the range 2–20 MHz, which may be compared with the auditory limit for humans, 20 kHz, and the upper limit of bat echolocation frequencies, 100 kHz. The speed of ultrasound in animal tissue is of order $1{,}500\,\mathrm{ms}^{-1}$ so its wavelength for a frequency of 10 MHz (speed ÷ frequency) is just 0.15 mm. We shall now consider how such frequencies can be generated, detected and used for imaging.

14.6.1. *The generation and detection of ultrasound*

The majority of ultrasonic generators are based on the phenomenon of *piezoelectricity*, the process by which a material placed in an electric field undergoes a mechanical strain and, conversely, when strained becomes electrically polarized and produces an electric field in its neighbourhood. The way that this happens depends on the way that atoms are bonded together. Figure 14.16(a) shows two atoms bonded together with electron clouds round the nuclei, each of which would be spherically symmetric for isolated atoms. The bonding results from displacement of the electron clouds to give a concentration of negative electron density between them so pulling inwards on the positively charged nuclei — which creates the bond. An imposed electric field, as shown in Fig. 14.16(b), would displace the electron clouds but if the outer electrons of atom B are more loosely bound, and hence move more easily, the result would be an increase of the negative charge density between the nuclei and hence they will be pulled closer together. Similarly, an electric field in the opposite direction will reduce the density between the nuclei and

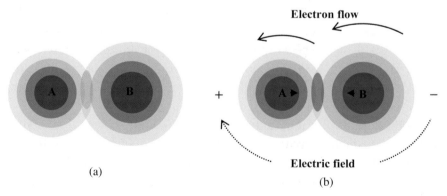

(a)

(b)

Figure 14.16   The electron density for (a) bonded atoms and (b) bonded atoms in an electric field.

they will move further apart. The net strain on the crystal, whether it expands or contracts, will depend on the aggregate effect of all the bonds within it. If the crystal has a centre of symmetry, i.e. for every pair of atoms AB there is a mirror image pair BA, then there will be no piezoelectric effect because an expansion due to one pair will be exactly balanced by a contraction due to the other.

The piezoelectric effect is reversible so that if a crystal is mechanically strained then the electron charge between the bonds will flow in and out of the regions between the nuclei, giving an electric polarization of the crystal and the generation of an electric field. Many common materials are piezoelectric and early devices usually used quartz crystals. Since the early 1950s the materials of choice are usually synthetic piezoelectric ceramics, such as barium titanate or, more recently, lead zirconate titanate. Although these materials are not crystalline it is still customary to use the term 'crystal' when referring to a piezoelectric component.

For the generation of an ultrasonic beam we require a transducer (§13.4) that converts electrical energy into ultrasound while, for detection, the conversion is from ultrasound to electrical energy. In both cases the transducer depends on the piezoelectric effect. The characteristic of the crystal that determines the frequency of sound it can generate is the natural frequency with which it vibrates. For a thin plate the main frequency with which it vibrates, by expansion and contraction perpendicular to the plate, corresponds to a wavelength of twice the thickness of the plate for sound waves within it. Thus if the speed of sound in the material of the plate is V and the thickness of the plate is d then the frequency with which it will vibrate, which will also be the frequency of any sound it generates, is

$$\nu = V/(2d). \tag{14.3}$$

The design of a basic ultrasound generator is shown in Fig. 14.17. The piezoelectric ceramic plate is sandwiched between two electrodes that apply an alternating field across it with a frequency equal to that of the natural vibration of the slice. The plate, which on its own would be very fragile, is mounted on a backing material. The design of the backing plate must ensure that the acoustic energy emitted by the

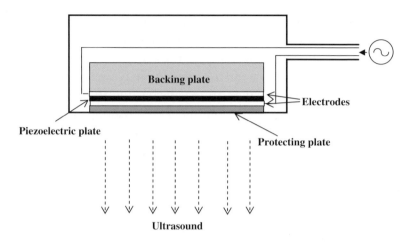

Figure 14.17   A piezoelectric ultrasound generator.

back surface of the piezoelectric element is efficiently absorbed and is not reflected back to interfere with radiation from the front surface. On the emitting side of the piezoelectric plate there is a layer of material, of quarter-wavelength thickness that increases the transmittance of the ultrasound, the principle being similar to that for lens blooming as described in §5.5. Finally there is a protection plate, usually made of steel, to prevent scratching of the blooming layer, which would give unwanted scattering of the emerging ultrasound.

The ultrasound that comes from the generator originates from all parts of the vibrating plate and is in the form of a gradually widening beam, rather like the beam from a hand-held torch. There are a number of different kinds of medical ultrasound procedure and what they all require is to bring the ultrasound to a focus to create a small well-defined high-intensity region, the reflection from which is to be measured. In modern ultrasound machines the focusing is done by phased-array technology in which a number of small ultrasonic generators are fed by the same source of alternating electric field but with different phase delays (§7.5.4). This enables not only focusing at different depths but also allows the direction of the ultrasound beam to be changed.

When a very short pulse of focused ultrasound is concentrated in a region of the body at the focus, then some ultrasound is reflected back towards the source. The transducer that produced the pulse is also able to detect the echo. The returning pulse sets the piezoelectric plate into vibration that generates an alternating field across the plate; the energy of the resultant current in the detecting circuit is proportional to the intensity of the echo. This information is digitized, stored in a computer and, when information to produce the complete image has been acquired, it can then be displayed on a screen.

14.6.2. *Medical ultrasonic procedures*

Ultrasound imaging can be performed in four different ways, depending on the type of information required. Very concentrated high-energy ultrasound applied over a considerable period is destructive to tissue so it can also be used in therapeutic mode to destroy malignant or non-malignant growths.

A mode

In this mode of operation, keeping the transducer fixed and varying the depth of focus, a single line through the body is explored. The intensity of the echoes plotted on a screen shows enhanced reflection at boundaries between different tissues and enables the thickness of organs to be measured in the direction of the scan.

B mode

A scan across the body is made by a linear array of transducers, all with a common depth of focus, which gives an image of a plane in the body. The resolution obtained with ultrasound, as for other types of imaging, is higher for higher frequencies but the penetrating power of ultrasound in tissue is better for lower frequencies. The choice of frequency must be a compromise between these two competing requirements. Because of its inherent safety as a procedure ultrasound scans are used for foetal monitoring to follow the progress

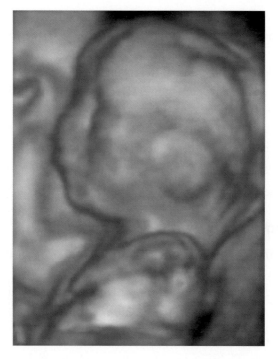

Figure 14.18   An ultrasound scan of a 20-week old foetus.

of the development of the foetus and to ensure that no problems are developing. An ultrasound scan image of a 20-week old foetus is shown in Fig. 14.18.

M mode

This produces a rapid sequence of B-mode scans and displays them on the screen in moving-picture form. This is particularly useful in studying the behaviour of the heart.

Doppler mode

The determination of the movement of fluids through the body, in particular blood, is of great importance in diagnosing some medical conditions. A partially blocked artery leading to impaired flow can be the precursor of a stroke or heart attack. A measure of the flow

Ultrasound

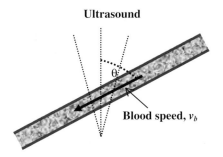

Blood speed, v_b

Figure 14.19   A schematic view of ultrasound falling on a blood vessel.

of blood is its speed of movement, which is not constant since, as the heart goes through its pumping cycle, the pressure, and hence blood speed changes. The reflection of ultrasound from blood is mainly from the red blood cells it contains. For the carotid artery, the one supplying oxygenated blood to the brain, the speed is normally in the range 0.03 to 0.18 ms^{-1}. This is a sufficiently large fraction of the speed of sound in bodily tissue to make Doppler shift a viable technique for measuring speed of flow. The situation when an ultrasound pulse falls on a blood vessel is illustrated in Fig. 14.19. The change of frequency of the returning pulse is given by Eq. (7.9) where we must use the radial speed component so giving

$$\frac{d\nu}{\nu} = \frac{2v_b \cos \theta}{V} \tag{14.4}$$

in which ν is the frequency of the ultrasound, $d\nu$ the Doppler shift in frequency, V the speed of sound in the tissue and the other symbols as shown in the figure.

As shown in Fig. 14.19, if the focus of the ultrasound is slightly beyond the blood vessel then there is a small volume of fairly high ultrasound density where both stationary tissue and moving blood are being irradiated. In the echo there would be two frequency components, the first corresponding to the stationary, or near stationary, tissue (the blood vessel walls will be dilating and contracting with differences of pressure) and the second corresponding to the moving blood. Since the ultrasound is contained within

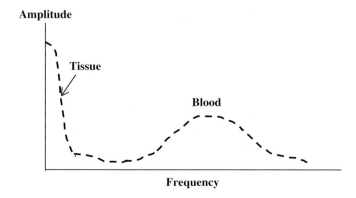

Figure 14.20   The frequency distribution from continuous ultrasound.

a narrow cone there will be small variations of θ that will give a spread of frequencies from the blood reflection. The appearance we might expect for the dependence of the amplitude of the returning signal with frequency is shown in Fig. 14.20. For the tissue the amplitude is high but the frequency spread low and for the blood signal the amplitude is lower and there is a spread of frequencies, partly due to the variation of θ but also due to variations of blood speed during the period of the pulse and within the sample volume.

There are two main types of Doppler-ultrasound procedure. The first uses a continuous wave of ultrasound. With the arrangement shown in Fig. 14.19 the returning signal will contain a range of frequencies as shown in Fig. 14.20. When converted into an electrical signal it can be analysed by electronic circuitry to give the spectral distribution and hence the distribution of speeds within the sample volume. This is a simple procedure and the equipment for carrying it out is inexpensive and portable.

The second procedure, using pulsed waves, requires a smaller sampling volume so that the spread of speed within it is small, giving a dominant frequency in the reflected pulse. The pulses are short, typically 6–40 wavelengths, which is of order 1 mm in total length. The technique used for measuring the Doppler frequency shift is similar to that described in §7.7.2 for pulsed-Doppler radar. Reference to

Fig. 7.20 is appropriate here. The output and reflected wave trains are stored and by mathematical means are overlapped. If one output pulse and reflected pulse are in phase then the next pair will have an appreciable phase difference and will partially cancel each other. If the phase difference were 180° they would completely annihilate each other. From the deduced phase difference the frequency change, and hence the speed of the material in the sample volume, can be found. There is the same problem with Doppler-ultrasound as there is with Doppler-radar. If the distance between the pulses is too large then the phase difference being measured will be greater than 180° and there will be an ambiguity in the frequency estimate. However, to measure the *distance* of the sample volume the interval between pulses cannot be too short, otherwise there will be an ambiguity in distance (§7.2). Sometimes a compromise pulse interval can be found that satisfies both requirements — otherwise the two properties, distance and speed, must be found separately using different pulse intervals. Figure 14.21 shows a typical spectrogram of the Doppler frequency variation with time in a small region within the carotid artery.

Although high intensity ultrasound can be destructive and is used for various medical procedures, such as the treatment of tumours or the destruction of kidney stones, under the conditions used for

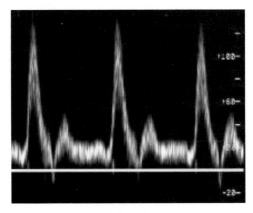

Figure 14.21   The variation of Doppler frequency (speed) with time within the carotid artery.

imaging it has no harmful side effects. For this reason it is widely used for examining the foetus during development, either to produce images of the whole foetus (Fig. 14.18) or for monitoring the foetal heart.

Chapter 15

Images of Atoms

15.1. The Nature of Crystals

Crystalline materials are commonplace in everyday life, ranging from household chemicals such as common salt, sugar and washing soda, industrial materials such as corundum and germanium to precious stones such as emeralds and diamonds. A characteristic of crystals is that they are bounded by facets — plane surfaces of mirror perfection. Plane surfaces are not very common in nature; they occur as the surfaces of liquids, but this is due to the influence of gravity on the liquid, a force that would not affect the form of a small rigid solid. It is easy to verify that the significance of the planar surfaces is not just confined to the surface of crystals but is also related to their internal structure. Many crystals can be cleaved along preferred directions to form new facets and even when a crystal is crushed into tiny fragments so that it seems to be a powder, examination through a microscope shows that each grain is a tiny crystal bounded by plane surfaces.

Another feature of crystals is that all the crystals of a particular substance tend to look alike — all plates or all needles, for example. The overall conclusion from these characteristics of crystals is that the arrangement of atoms within a crystal governs its overall appearance. The flatness of crystal facets can be interpreted as due to the presence of planar layer of atoms forming a surface of low energy and hence one that is intrinsically stable.

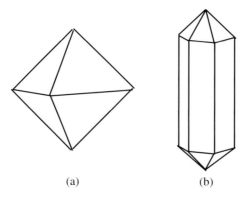

(a)   (b)

Figure 15.1   (a) An alum crystal. (b) A quartz crystal.

15.1.1. *The shapes of crystals*

Many crystals are very regular in shape and show a great deal of symmetry. Figure 15.1 shows two very regular shapes, an alum crystal with the form of an octahedron (a solid with surface a symmetrical arrangement of eight equilateral triangles) and a quartz crystal with a cross-section that is a perfect hexagon.

Although most alum crystals seem alike, they are not all perfect octagons but may appear to be a distorted form of an octagon. For a two-dimensional example we take the form of platy crystals of a particular substance as shown in Fig. 15.2. They have the common feature that they have six sides but they are all very different in appearance. Now we take a point within each of the plates and draw arrows from it perpendicular (normal) to the sides of the plate. In each case the set of arrows is the same and is what would be obtained from a regular hexagon, although none of the platy crystals has the form of a regular hexagon. Thus, although the plates are all different in appearance, the fact they all show in the normals to their faces the characteristic of a regular hexagon reveals that there is a hexagonal symmetry in the underlying atomic structure that forms the crystal.

What we have found for a two-dimensional example also applies in three dimensions. Figure 15.3(a) shows a set of normals for an octahedron and the solid shown in Fig. 15.3(b) has the same set of

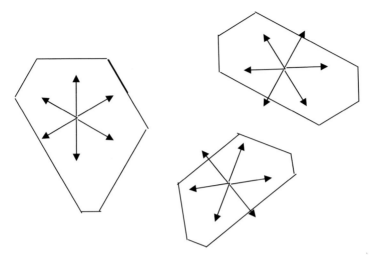

Figure 15.2   Three platy crystals, all showing a basic hexagonal symmetry.

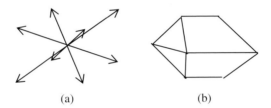

(a)   (b)

Figure 15.3   (a) A set of normals to the faces of an octahedron. (b) A solid with the same set of normals.

normals; although it is not an octahedron it could be the form of an alum crystal.

15.1.2. *The arrangement of atoms in crystals*

Crystals consist of three-dimensional periodic arrays of atoms. That accurate and succinct statement is perfectly true but may not give an immediate appreciation of the structure of a crystal at the atomic level. So, to give a more visual image, Fig. 15.4 shows a schematic two-dimensional crystal containing three different kinds of atom, indicated by different shading. It consists of a basic unit in the form of a parallelogram, shown in heavy outline and called a *unit cell* by

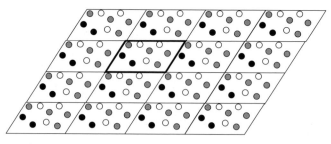

Figure 15.4   A two-dimensional crystal. The unit cell is indicated in heavy outline.

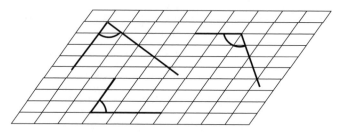

Figure 15.5   Three possible angles between facets for a two-dimensional crystal.

crystallographers, packed side by side in two dimensions to create the whole crystal. In a real three-dimensional crystal the unit cell is a parallelepiped with the atoms arranged within its volume. Packing these units in three dimensions then forms the complete crystal. Although the two-dimensional figure is limited to a 4 × 4 array of unit cells, a real crystal would usually have several million cells in each of the three packing directions.

Long before any scientific tools were available to explore crystals in atomic detail, mineralogists were deducing the shapes of unit cells by measuring the angles between the facets of crystals with optical instruments called *goniometers*. The general principle for finding shapes is illustrated in two dimensions in Fig. 15.5, which shows how various angles could come about. The heavy lines are possible facets formed by the arrangements of unit cells at the boundary of the crystal. The angles shown could only come about by having a unit cell with a particular angle and sides with a particular *ratio* of lengths. Only the shape of the unit cell and the relative sizes of the unit cell edges for a crystal could be determined by the technique of measuring angles between facets, known as *optical goniometry*; the determination of the actual sizes of the cell edges had to await more advanced physical techniques.

15.2. The Phenomenon of Diffraction

In Chapter 3 the wave nature of light was described. Light is scattered when it impinges on an object; a lens may be used to recombine the scattered light into an image or, in the case of a hologram, the scattered light can be combined with a reference beam and recorded on a photographic plate. Here we are going to consider another aspect of light scattering, the behaviour of a diffraction grating.

15.2.1. *A one-dimensional diffraction grating*

The standard form of a simple one-dimensional diffraction grating is an array of equally spaced transparent slits set in an opaque background. Monochromatic light falling normally on this array of slits will give strong beams for directions in which the light from all the slits is in phase, i.e. directions for which the path difference in the light coming from neighbouring slits is a whole number of wavelengths. Examples of such directions are shown in Fig. 15.6 where strong diffracted beams are produced in directions corresponding to $n = 0, \pm 1, \pm 2$, where the light scattered by each slit has a path difference $n\lambda$ with that from a neighbouring slit.

The diffraction grating shown in Fig. 15.6 has been described as *simple* because there is just one scattering object — the slit — in each interval. The intensities of the diffracted beams will not be equal, because scattering tends to be stronger in the forward direction and fall-off on either side, but the fall-off would be predictable and gradual as one moves outwards from the central peak.

It is possible to have a more structured diffraction grating and one such is shown in Fig. 15.7(a). Now the periodic unit of the grating contains not one slit but three, all of different widths and hence of different scattering power.

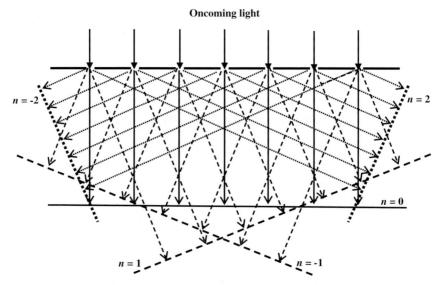

Figure 15.6   The wave fronts of diffracted beams from a simple one-dimensional diffraction grating.

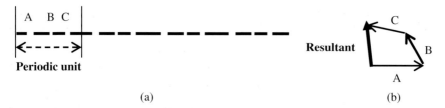

Figure 15.7   (a) A structured diffraction grating. (b) The resultant for one of the diffracted beams.

All the A slits alone will give strong diffracted beams as shown in Fig. 15.6, as will the slits B alone and the slits C alone. However, there will be phase differences between the contributions of the three sets of slits and the way that these contributions combine to give the amplitude of the resultant is shown in Fig. 15.7(b). The intensity of the diffracted beam is proportional to the square of the resultant amplitude. The difference between the intensity patterns of the simple single slit grating and that for a complex structure, one containing many slits, in a case with many orders of diffraction (values of n) is

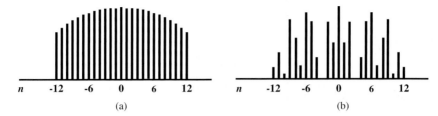

Figure 15.8   The intensities of diffracted beams for (a) a grating with a single scatterer per repeat interval and (b) a grating with many scatterers per repeat interval.

illustrated in Fig. 15.8. The gradual fall-off of intensity for the simple grating contrasts with the apparently irregular variation for the complex grating. This variation of intensities comes about because of the arrangement of scattering slits and their different scattering powers and, in principle, from the intensities the arrangement of slits could be deduced.

15.2.2. *A two-dimensional diffraction grating*

A two-dimensional diffraction grating of a simple kind, consisting of a network of small circular apertures on an opaque background is illustrated in Fig. 15.9(a). The diffraction pattern when light is shone normally onto the grating is shown in Fig. 15.9(b).

The grating and its diffraction pattern are both rectangular arrays and it will be seen that the spacing of the diffracted beams is closer along x, in which direction the spacing of the apertures in the grating is greater. Theory shows that the product of the two spacings is the same along the x and y directions, so that

$$a \times a^* = b \times b^* \tag{15.1}$$

and from the values of a^* and b^*, together with a knowledge of the wavelength, λ, of the radiation being used for diffraction, the values of a and b can be found. In the event that the original grating is based on an oblique, not rectangular, lattice then the diffraction pattern is also on an oblique lattice and from a^*, b^*, the angle between a^* and b^* and λ the size and shape of the grating lattice can be found.

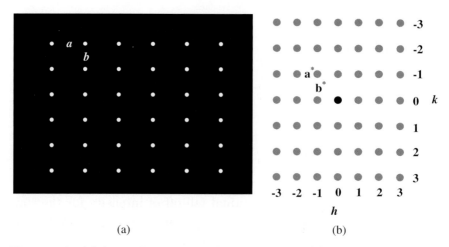

(a)   (b)

Figure 15.9   (a) A two-dimensional diffraction grating. (b) The two-dimensional diffraction pattern.

For a one-dimensional lattice the orders of diffraction were defined by integers n and for the two-dimensional lattice by two sets of integers, indicated as h and k in Fig. 15.9(b). For the simple two-dimensional lattice, with one scatterer per unit, the intensities would gradually fall off with distance from the central beam ($h = 0$, $k = 0$) because of the fall-off of scattering amplitude with angle of scatter. However, if each lattice unit contained a number of scatterers, similar to what is shown in Fig. 15.4, then there would be irregular variations of the intensities of the various scattered beams, caused by the distribution of scatterers in each unit and their relative scattering powers. Just as for the one-dimensional case, the variations of intensity for the different pairs of values of h and k could, in principle, reveal the positions and scattering powers of the contents of each identical unit cell.

15.3. The Beginning of X-ray Crystallography

In 1912, during a walk in the Englischer Garten in Munich, the German physicist Max von Laue (1879–1960; Fig. 15.11(a)) was talking to a young student, Paul Ewald (1888–1985), about his doctoral thesis, which was concerned with the interaction of light

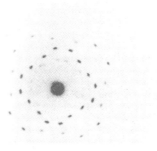

Figure 15.10   A diffraction pattern of zincblende obtained by Friedrich and Knipping.

with crystals. At that time it was known that x-rays are an electromagnetic radiation with wavelengths of order 10^{-10} m and von Laue suddenly realised that crystals were three-dimensional periodic objects of about the right size to act as diffraction gratings for x-rays. He asked two researchers in his laboratory, Walter Friedrich (1883–1968) and Paul Knipping (1883–1935), to look into this possibility, so they set up an experiment in which a beam of x-rays was fired at a crystal of zincblende (zinc sulphide) with a photographic plate placed to record any scattered radiation in a broadly forward direction. Even though the quality of the image was poor, their first results showed quite clearly the presence of spots on the developed plate, which indicated that diffraction had occurred. A later experiment gave a better result, shown in Fig. 15.10. Von Laue, Friedrich and Knipping published a scientific paper announcing their result and offering an explanation for the arrangement of spots they had observed.

The paper attracted a great deal of attention, in particular that of a young Australian-born British first-year research student at Cambridge, William Lawrence Bragg (1890–1971; Fig. 15.11(b)), generally known as Lawrence Bragg. He deduced that the theoretical explanation of the diffraction pattern given in the paper was wrong and he found that it was possible to interpret the pattern in terms of mirror-like reflections from planes in the crystal that were richly populated with atoms. The presence of such planes is illustrated in two-dimensions in Fig. 15.12. There are lines that pass through atoms

(a)   (b)   (c)

Figure 15.11   Three pioneers of x-ray crystallography: (a) Max von Laue; (b) William Lawrence Bragg; (c) William Henry Bragg.

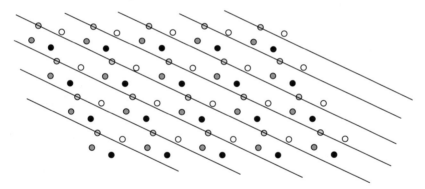

Figure 15.12   A hypothetical two-dimensional crystal showing richly populated lines.

indicated by the different shades of grey parallel to the ones shown for the pale grey atoms. In three dimensions there will be many planes containing similar atoms from all the unit cells.

William Henry Bragg (1862–1942; Fig. 15.11(c)), Lawrence Bragg's father, was Professor of Physics at Leeds University and, for another purpose, had previously constructed an x-ray spectrometer that could be used precisely to measure the wavelengths of x-rays and also the intensities of the diffracted beams. He and his son began to work together to collect diffraction patterns from the crystals of

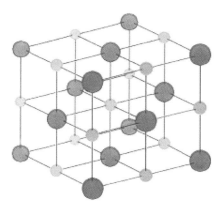

Figure 15.13   The crystal structure of sodium chloride. Large circle = sodium, small circle = chlorine.

very simple structures, such as common salt, sodium chloride (NaCl), and to deduce their crystal structures, i.e. to find the size and shape of the unit cell and the positions of atoms within the unit cell. They used fairly ad hoc trial-and-error methods to find atomic positions that would give the observed diffracted-beam intensities; the structure they found for sodium chloride is shown in Fig. 15.13. For this pioneering work von Laue received the Nobel Prize for Physics in 1914 and the two Braggs the Nobel Prize for Chemistry in 1915. Lawrence Bragg heard of his prize while serving as an artillery officer in France during the First World War. On the other side of the front line Paul Ewald, who was the stimulus for this new subject of x-ray crystallography, was manning a mobile x-ray unit, helping in the treatment of wounded German soldiers. It is pleasing to record that as scientists working in the same field these two pioneers later established a strong friendship.

15.4. X-rays for Diffraction Experiments

In §14.2, in which x-ray generators were described, we gave no thought to the detailed wavelength distribution of the x-rays being produced. However, it turns out that diffraction experiments from crystals are best carried out with monochromatic x-rays so we now turn our attention to how these may effectively be produced.

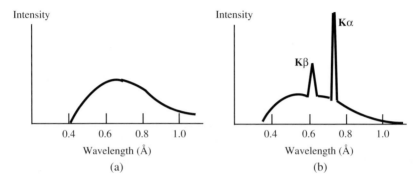

Figure 15.14   Output from a molybdenum tube at (a) 30,000 V and (b) 35,000 V.

The basic form of an x-ray tube for generating radiation for crystallographic experiments is essentially as shown in Fig. 14.4, except that the anode material may be of copper (Cu), molybdenum (Mo), chromium (Cr) or some metal other than the tungsten usually used for medical purposes. The intensity distribution of the emitted x-rays depends on the target material and the potential difference across the tube. In Fig. 15.14(a) we show the output from a molybdenum tube with a potential difference of 30,000 V. There is a smooth variation of intensity with wavelength. Electrons impinging on the anode lose some of their energy in producing heat, removed by water-cooling, and the remainder in one or more interactions with molybdenum nuclei in producing x-rays. The most energetic x-rays that can be produced are those with energy 30,000 eV, which is when an electron loses all its energy in a single collision. This corresponds to a wavelength of 4.1×10^{-11} m $= 0.41$ Å, the short wavelength cut-off seen in the figure. Reducing the potential difference would both give lower intensities and also move the cut-off to a longer wavelength. The radiation produced in this way is known as *bremsstrahlung*.

If the tube potential is raised to 35,000 V then a new feature appears in the intensity curve — spikes of high intensity (Fig. 15.14(b)) referred to as *characteristic radiation*. At this energy the impinging electrons have enough energy to eject electrons from the innermost shell of electrons surrounding the molybdenum nucleus — the K shell. Outside this there are electrons in other shells,

L, M, etc., with successively higher energy with progress through the alphabet. An electron from a higher shell quickly fills the vacancy in the K shell and when this happens the electron gives up the energy difference between its original and new states in the form of an x-ray photon. The most probable transition is from the L shell, which gives the intense Mo Kα peak. Actually there are two slightly different energy levels in the L shell from which the electron can come so there are two unresolved peaks, Mo Kα_1 and Mo Kα_2 at 0.70926 Å and 0.71354 Å, respectively. A less common transition is from the M shell, giving the Mo Kβ peak at 0.6323 Å.

Most x-ray crystallographic measurements are carried out with monochromatic x-rays and the intense Kα radiation is obviously a good potential source if other wavelengths can be removed — in particular the less intense but still strong Kβ radiation. A simple way of doing this is with a *filter*, a thin metal foil through which the x-rays pass. When an x-ray photon passes through material it is absorbed and, in general, the shorter is the wavelength the greater is the penetrating power and the less is the absorption. Figure 15.15 shows the absorption as a function of wavelength for a typical metal. Starting with wavelength corresponding to point A, as the wavelength is shortened, so the x-rays become more penetrating and the absorption reduces. However, at the wavelength corresponding to point B the electron has enough energy to eject one of the electrons in the L shell. When this new way of taking up the photon energy becomes

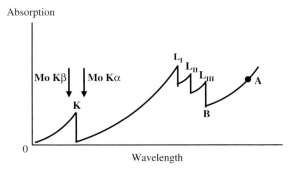

Figure 15.15   The variation of absorption with wavelength for a typical filter material.

Table 15.1 Some target materials and their Kα wavelengths.

Anode material	Kα wavelength (Å)
Silver	0.5594
Molybdenum	0.7092
Copper	1.5405
Nickel	1.6578
Cobalt	1.7889
Iron	1.9360
Chromium	2.2896

available the absorption sharply rises; at this wavelength there is an *absorption edge*. With further shortening of the wavelength the absorption again begins to fall until the next absorption edge is met. The complete absorption diagram is shown in the figure. Now imagine that the x-ray photons are coming from a molybdenum tube and that Mo Kα and Mo Kβ straddle the K absorption edge of the filter, as shown. The Mo Kβ radiation and the bremsstrahlung are quite heavily absorbed whereas there is little absorption of Mo Kα. In this way, by the use of a zirconium (Zr) filter, virtually monochromatic Mo Kα radiation can be produced. Similarly a nickel (Ni) filter can be used to produce almost monochromatic Cu Kα radiation.

A selection of useful anode target materials and their Kα wavelengths are given in Table 15.1.

If the intensity of radiation emitted by an x-ray generator tube is increased by increasing the heating of the cathode, and hence the flow of electrons, there can be problems with cooling the anode, which is intensely heated over a small region. A way of overcoming this difficulty is with a *rotating anode tube* in which the anode is spun in such a way that different parts of the anode are exposed to the electrons as it rotates. In this way much more intense x-ray beams are possible. However, the ultimate in x-ray beam intensity is obtained from *synchrotrons*, large and expensive machines provided centrally for the scientific community. A schematic synchrotron is shown in Fig. 15.16.

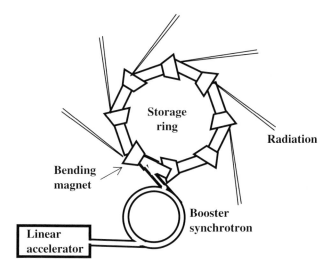

Figure 15.16   A typical x-ray synchrotron source.

The linear accelerator produces electrons of energy about 10 MeV that are injected into the booster synchrotron, which accelerates the electrons round a circular path until their energy reaches about 500 MeV at which stage they are injected into the main storage ring. Here the electron energy is gradually ramped up to the region of 2–4 GeV[1] by being passed through radio-frequency cavities. They traverse a polygonal path by having their paths deflected by dipole bending magnets. Eventually the electrons are isolated in the ring and repeatedly travel round the ring, typically for periods of several hours before they need to be refreshed. Wherever the magnets bend the path of the electrons they are in a state of acceleration; the laws of physics indicate that accelerating electrons emit radiation, which they do in a narrow beam. The radiation usually covers a wide range of wavelengths, from fractions to several hundreds of Ångstrom units. A characteristic spectrum of the output is shown in Fig. 15.17. This outpouring of radiation is at the expense of the energy of the electrons so, to maintain the current in the ring, energy is constantly fed in through the radio-frequency cavities.

---

[1] $1 \, \text{GeV} = 10^9 \, \text{eV}$.

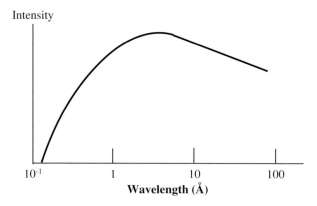

Figure 15.17   A typical synchrotron output spectrum.

To give an idea of the scale of a modern synchrotron, the Diamond Light Source, completed in the UK in 2007, is 561 m in circumference and has 32 bending magnets. It produces radiation from x-rays through to the infrared; the actual output can be focused and tuned to particular wavelengths by *insertion devices*, called *wigglers* and *undulators*. The brightness of the source is 10^8 times that of the Sun. For the most demanding work these are now the sources of choice of x-ray crystallographers.

15.5. The Phase Problem in Crystallography

Because of Lawrence Bragg's way of describing the phenomenon of x-ray diffraction, it is customary to refer to the individual diffracted beams as *reflections*, each associated with a triplet of integers (h, k, l), called the *Miller indices*, and an intensity $I(h, k, l)$. The angle made by the incident beam to the normal to the reflecting plane is conventionally indicated by θ, so the reflected beam makes an angle 2θ with the incident beam. The arrangement of atoms in a crystal usually, but not always, displays a great deal of symmetry — such as a *mirror plane* in which each atom has a partner mirrored in the plane. Another kind of symmetry that may occur is an *axis of symmetry*; a threefold symmetry axis indicates that looking down the axis one would see an equilateral triangle of identical atoms all in the same plane perpendicular to the axis. There are a large number of possible *symmetry elements*, many quite complicated to describe, but we need not concern ourselves with them for present purposes. Suffice to say that there are 230 distinctive combinations of symmetry elements, each combination referred to as a *space group*. The pattern of reflections, in both position and intensity, often reveals uniquely the space group, or at least limits the choice of space group, before the actual crystal structure has been solved.

The individual intensities depend on both the Miller indices and the types and positions of atoms in the unit cell. The unit cells are in the form of a parallelepiped (Fig. 15.18), defined by the cell edges, of lengths a, b and c and the angles α, β and γ shown in the figure. The directions of a, b and c define a set of, not necessarily orthogonal, Cartesian axes. Measured from one corner of the unit cell as an origin, the positions of the atoms are defined by their *fractional coordinates*. Thus, if the actual position of an atom is given by the coordinates (X, Y, Z) then the fractional coordinates are $(x, y, z) = (X/a, Y/b, Z/c)$.

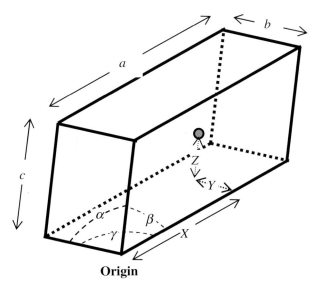

Origin

Figure 15.18   A unit cell, showing cell dimensions, an atom and its coordinates (X, Y, Z).

The scattering of x-rays by an atom is almost completely due to its electrons that, for the purpose of x-ray diffraction, can be thought of as a spherically symmetric cloud of electron density surrounding the nucleus. Thus the heavier the atom, i.e. the more electrons there are associated with it, the greater will be its scattering power, which is greatest in the forward direction in which it is proportional to the number of electrons in the atom — its *atomic number*. Due to inter-ference between scattered radiation coming from different parts of the electron cloud the scattering power falls of with angle of scatter-ing, 2θ, which is the angle a particular reflected beam makes with the incident beam direction. The fall-off also depends on the wave-length of the x-rays, λ, because the smaller is λ the greater is the phase difference, and hence interference, between scattering from dif-ferent parts of the electron cloud. When the mathematics is done it is found that the scattering power, defined as the *scattering factor*, of an atom is a function of $\sin\theta/\lambda$. The scattering factors for a few atoms are shown in Fig. 15.19; they are expressed in units of the scattering from a hypothetical point electron, which would scatter equally in all directions.

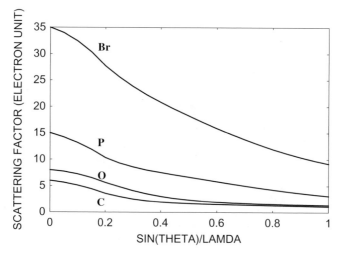

Figure 15.19   The scattering factors for carbon (C), oxygen (O), phosphorus (P) and bromine (Br) in electron units.

To find the intensity of an x-ray reflection in terms of the scattering factors of the atoms and their positions in the unit cell, which takes into account the way that the contributions of different atoms combine (Fig. 15.7), we must first define another quantity, the *structure factor*, given by

$$F(h, k, l) = \sum_{j=1}^{N} f_i \exp\{2\pi i (h x_j + k y_j + l z_j)\} \qquad (15.2)$$

where the sum is over the N atoms in the unit cell, the jth one of which has scattering factor f_j for the reflection of indices (h, k, l) and fractional coordinates (x_j, y_j, z_j). Mathematically

$$\exp\{2\pi i (h x_j + k y_j + l z_{ij})\} \equiv \cos\{2\pi i (h x_j + k y_j + l z_j)\}$$
$$+ i \sin\{2\pi i (h x_j + k y_j + l z_j)\} \qquad (15.3)$$

$F(h, k, l)$ has real and imaginary components, so it is a complex quantity — a sum of real and imaginary parts — which may be expressed in modulus-amplitude form as

$$F(h, k, l) = |F(h, k, l)| \exp\{i\phi(h, k, l)\} \qquad (15.4)$$

where $|F(h, k, l)|$ is called the *structure amplitude* and $\phi(h, k, l)$ the *phase* of the reflection. The intensity of the reflection (h, k, l) is then proportional to $|F(h, k, l)|^2$; there are other factors contributing to the intensity due to the geometry of the diffraction experiment and absorption by the crystal but these can be determined and corrections made for them.

This brief mathematical account shows that the intensity may be expressed in terms of the structure but the problem facing the x-ray crystallographer is the inverse of this; the intensities can be measured but what is required is to find the structure. This is determined in the form of mapping the electron density in the crystal. Each atom appears in such a map as a blob of electron density with the atom situated at the central point of highest density. This density, $\rho(x, y, z)$ at the point with fractional coordinates (x, y, z), can be calculated from

$$\rho(x, y, z) = \sum_{\text{All reflections}} |F(h, k, l)|$$

$$\times \cos\{2\pi(hx + ky + lz) - \phi(h, k, l)\} \qquad (15.5)$$

This equation summarizes the major problem faced by the x-ray crystallographer. The structure amplitude, $|F(h, k, l)|$ may be found from the diffraction experiment — it is just proportional to the square-root of the intensity — but the phase does not come from the experiment. This is the *phase problem in crystallography.*

15.6. Determining Crystal Structures; Electron-density Images

Many techniques have been devised to solve the phase problem.[2] The very first structures solved by the Braggs, and those that closely followed them, were very simple ones containing few atoms, which could be moved around in the unit cell in a trial-and-error process until a match with the intensities was found. Later, more advanced techniques were developed but, even so, for many years only simple structures with up to forty or so atoms in the unit cell could be solved. The first protein structures were solved in the late 1950s by the Cambridge scientists, John Kendrew (1917–1997), who solved *myoglobin*, and Max Perutz (1914–2002), who solved *haemoglobin*; they shared the 1962 Nobel Prize for Chemistry for this pioneering work. As methods improved, so structures of ever-increasing complexity have been solved, including those of some viruses. Depending on the type of structure, the x-ray maps, which are the calculated images of atoms within crystals, can be interpreted either as frameworks, where the bonding is continuous throughout the crystal, or as molecular structures, where separated tightly bonded groups of atoms form discrete entities. The example of a framework structure

---

[2]Details of methods for solving crystal structure may be found in Woolfson, M.M. and Fan Hai-Fu (1995) *Physical and Non-physical Methods of Solving Crystal Structures*, Cambridge, Cambridge University Press.

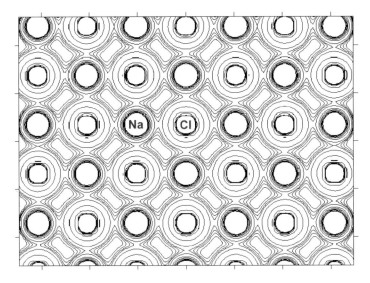

Figure 15.20   A section of an electron-density map for sodium chloride.

electron-density map, seen in Fig. 15.20, is for sodium chloride, NaCl. In this ionic structure a single electron transfers from sodium atoms to chlorine atoms thus giving each type of atom a completely filled outer shell of electrons. This confers a positive charge on the sodium atom, because of the lost electron, and a negative charge on the chlorine atom, because of the gained electron. The resultant attraction between oppositely charged neighbouring atoms binds the lattice structure together with *ionic bonds*. After the transfer of electrons there are 10 associated with each sodium atom and 18 with each chlorine atom; it will be seen in this section of the electron-density map that there is a greater concentration of contours around chlorine.

At a different level of complexity Fig. 15.21 shows part of the projected electron density for a small protein *lysozyme*, which is present in egg white, in saliva and in tears and is an anti-bacterial agent, breaking down the cell walls of bacteria. Because hydrogen has a single electron — and that well spread out — it is normally not picked up in electron-density maps unless very precise data are obtained. The lysozyme structure contains 1,001 non-hydrogen

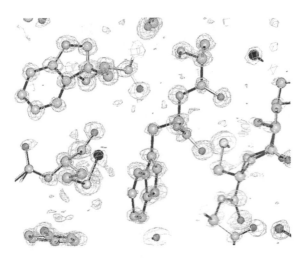

Figure 15.21   Part of the projected electron density for the protein lysozyme.

atoms, including 10 sulphur atoms; in the figure lines are drawn to show chemically bonded atoms.

Although electron-density maps can be obtained for very large structures, such as viruses, they are so complicated that it is difficult to interpret what they mean. For this reason it is customary to present the information in the maps of large proteins and viruses in a simpler form, showing major structural features. Figure 15.22 shows part of the solved structure of tobacco mosaic virus, which attacks a variety of plants, including tobacco, causing marking on the leaves — hence its name. In the figure the helical shapes are *α-helices*, chains of amino-acids forming variable lengths of helical structure. The thin wandering lines are just protein chains forming no particular characteristic shapes.

X-ray crystallography was the means by which the British scientist Francis Crick (1916–2004) and the American scientist James Watson (b. 1928), who together with Maurice Wilkins (1916–2004) were awarded the Nobel Prize for Physiology or Medicine in 1962, solved the structure of deoxyribonucleic acid (DNA) — possibly the greatest scientific breakthrough in the twentieth century, which laid the foundation for the age of biotechnology.

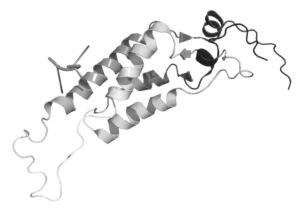

Figure 15.22   A schematic representation of the structure of the coat protein of tobacco mosaic virus.

15.7. The Scanning Tunnelling Microscope

In the bulk of a crystalline solid each atom is bonded to a number of surrounding atoms. The physical and chemical properties of the bulk solid are related to this system of bonds so that, for example, stretching the solid may be easier in some directions than others so that the elasticity of the solid is non-isotropic. At the surface of the solid, or at the bonded interface between adjacent different materials, the arrangement of atoms and the way they are bonded must be different from that in the bulk material if the surface or interface atoms are to be in an equilibrium configuration; it follows that the properties in the surface region will also be different. Sometimes the properties of surfaces and interfaces can be exploited for practical purposes, so knowledge of the atomic structure of surfaces is of interest to scientists and engineers. There are many ways of exploring surface structure but here we shall just consider one of these — the scanning tunnelling microscope (STM) — which gives clear imaging of the atomic arrangement in surfaces.

The phenomenon of *tunnelling* is one that is non-classical and is explained by quantum mechanics. In Fig. 15.23 an electron is shown moving along the x-axis with kinetic energy K and zero potential energy. Hence its total energy E, kinetic + potential, is K and if it

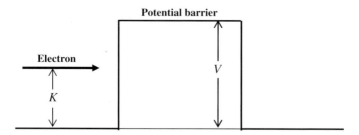

Figure 15.23   An electron approaching a potential barrier.

moves into an area in which its potential energy changes then the kinetic energy changes to maintain the total energy constant. In the figure we see that the electron meets a potential barrier, within which the potential energy $V > E$. To meet the condition that the sum of kinetic and potential energy must equal E the kinetic energy would have to be negative, which is impossible, so the electron cannot enter the barrier and according to classical theory it is reflected with the original kinetic energy, K.

When the situation portrayed in Fig. 15.23 is analysed by quantum mechanics, which gives the electron a wave-like nature, the result is that there is a finite probability that the electron will go through the barrier and emerge on the other side with the same kinetic energy as that with which it entered. Of course, each individual electron either goes through the barrier or is reflected in the way predicted by classical mechanics, so what the quantum mechanical calculation gives is the proportion of electrons that gets through. As intuition indicates, the proportion getting through reduces as the barrier gets higher and also as it gets thicker.

Now we consider the situation shown in Fig. 15.24 in which two metallic plates are connected respectively to the two sides of a battery but are separated. If the separation is large then no current will flow. If the two plates are brought closer and closer together then classical theory tells us that no current will flow until they actually touch. The finite gap acts like a potential barrier to the flow of electrons. However, quantum mechanics tells us a different story — that when the gap becomes *very* small some electrons will be able to get through

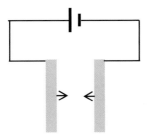

Figure 15.24   Two separated metal plates with a potential difference between them.

the gap. The smaller the gap the larger will be the current, and the magnitude of the current is sensitive to small changes of gap.

With a potential difference of a few volts between the plates the maximum distance apart of the plates for obtaining a measurable current is less than 1 nm — so very tiny indeed. In 1981 the German physicist Gerd Binnig (b. 1947) and the Swiss physicist Heinrich Rohrer (b. 1933) working at the IBM Research Laboratory in Zurich developed the STM, a device for looking at the atomic structure of surfaces; for this work they jointly received the Nobel Prize for Physics in 1986. The principle of the device is very simple — but the technical demands are huge. The principle is illustrated in Fig. 15.25. A very fine metallic pointer, with an effective tip end of width less than the diameter of an atom, is moved over the surface of a specimen at a distance typically in the range of 4–7 Å. The specimen material has to be conducting and a potential difference of about 4 V is maintained between the pointer and the surface. Because of the approximately spherical shape of the atoms (actually the electron clouds that form their volume), as the pointer moves over the surface the effective gap from pointer tip to surface changes. Although the changes in the gap are small in absolute terms, they are quite substantial in relation to the size of the gap and, as has been mentioned, the current is very sensitive to the gap distance.

It can be seen that the principle of the STM is straightforward but putting that principle into practice presents a difficult problem, which is the one that Binnig and Rohrer solved. The two major problems are, firstly, producing a pointer with an effective tip of width

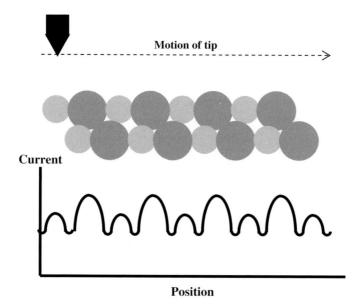

Figure 15.25   The basic operation of an STM.

less than that of an atom and then, secondly, moving that tip with a precision of a fraction of an Ångstrom unit.

Producing a pointer with a fine tip is the most difficult part of the instrumentation of an STM. These are normally made of tungsten, gold or an alloy of platinum and iridium. Ideally there should be a single atom at the end of the pointer through which the current flows to the specimen and when this is achieved the image will give atomic resolution. Very high precision grinding has been used to produce a fine tip and other processes such as chemical etching have been used. Another way of producing a fine tip is to raise the voltage between tip and sample to about 7 V, which causes tungsten atoms to migrate towards the tip, leaving a single atom at the apex. An ideal tip end in terms of its atomic structure is shown in Fig. 15.26.

The fine motion of the point, which must be carried out in three dimensions, two for a raster scan of a small region of the specimen and the third to control the distance from the specimen, is implemented using the piezoelectric effect (§14.6.1). In Fig. 15.27 a slab of piezoelectric material of thickness z is shown with a potential difference

Figure 15.26   An ideal tip of an STM pointer.

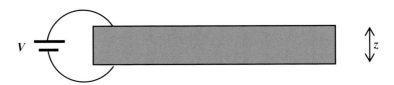

Figure 15.27   A piezoelectric material with a potential difference across it.

V across it, producing an extension ε. The strain induced in the material, defined as extension divided by thickness is

$$s = \varepsilon/z \qquad (15.5\text{a})$$

and the field across the slab is

$$E = V/z \qquad (15.5\text{b})$$

The piezoelectric constant of the material, d, that, in general, is non-isotropic is given by the strain per unit field so that

$$d = \frac{\varepsilon/z}{V/z} = \varepsilon/V \qquad (15.5\text{c})$$

or

$$\varepsilon = dV. \qquad (15.5\text{d})$$

Typical values for the piezoelectric constant d, which can be positive or negative, have magnitudes in the range 1 to 6 $\text{Å}\text{V}^{-1}$ so that very tiny movements can be produced by voltage changes that can be accurately controlled and measured.

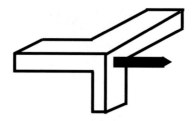

Figure 15.28   An STM pointer with three orthogonal piezoelectric bars for positional control.

There are many possible designs for piezoelectric positioning devices. The simplest is shown in Fig. 15.28 that consists of three orthogonal bars of piezoelectric material with potential differences along their lengths controlling their tiny extensions and contractions.

To use an STM the specimen must be mounted on a horizontal platform, well isolated from vibrations; since measurements are being made with a precision that is a fraction of the size of an atom it is essential that no parts of the equipment experience extraneous movement. The pointer is then brought close to the surface by a coarse control, which may be of various forms but typically could be a fine screw, the rotation of which raises and lowers the tip. The voltage on the piezoelectric device controlling motion perpendicular to the specimen surface (the z direction) is used make the final adjustment; the level at which to set the pointer is indicated by the current flowing between pointer and specimen, which is of order a few milliamps. There are then two major ways of operating the STM.

Constant level

The pointer is moved in raster fashion over a small rectangular region of the specimen, a few tens of Ångstroms in each dimension without changing the position of the pointer in the z direction. The variation of current then indicates the differences of distance of the specimen relative to the tip. This is stored in a computer memory and displayed on a screen, often with false colour added to help interpretation of the image.

Figure 15.29   An STM image of the [0 0 1] plane of the gold lattice.

Constant current

In this mode as the raster motion is performed the height of the pointer is adjusted to maintain a constant current between tip and specimen, meaning that the z distance is constant. The voltage on the piezoelectric controller for the *z direction* is recorded and used to create the image on a computer screen.

Figure 15.29 shows an STM image of a surface of gold atoms. Each individual atom can be clearly seen as a bright dot and, for this particular plane in the gold lattice, there is also a corrugation of the surface, each accommodating five rows of gold atoms.

Figure 15.30(a) shows two wire models of carbon nanotubes, equivalent to plane hexagonal graphite structure sheets bent into a cylindrical surface and joined seamlessly in two different ways. There are several forms of carbon nanotube, depending on how the pattern of hexagons fits together, but they all exhibit remarkable physical properties. The hexagonal structures are extremely rigid and their resistance to stretching is five times higher than that of steel. They also have a tensile strength some 50 times that of steel that, combined with their low density, make them good candidates for use in the aerospace industry. An STM image of a carbon nanotube

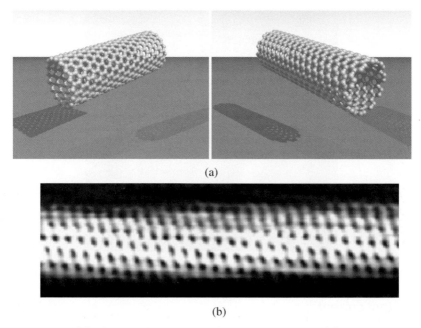

(a)

(b)

Figure 15.30   (a) Models of two forms of carbon nanotube. (b) An STM image
of a carbon nanotube.

(Fig. 15.30(b)) clearly shows the characteristic structure where each
bright spot is an individual carbon atom.

The STM is a wonderful example of a precision technology that
is leading to a new understanding of the structure of materials at an
atomic level. Although this may seem to be the limit in the small
scale of entities that we can detect and image this is not so. Small
as they are, atoms are built of even smaller particles, which cannot
be imaged as such but whose motions can be imaged — the topic of
our final chapter.

Chapter 16

Images of Particles

The term imaging is generally understood to be the creation of an image of an object by means of recombining the radiation, or other type of wave motion, that it has scattered. In previous chapters, when direct imaging has been impossible or extremely difficult, we have stretched that definition to include the creation of a visible image by some technical process — computation in the case of x-ray crystallography. Many of the imaging processes we have considered have involved electromagnetic radiation outside the visible range or wave motions that are not electromagnetic radiation, e.g. sound waves. These have all been subject to the rule for seeing the structural detail of a viewed object — that the wavelength of the irradiating waves should be of the same order as the size of the detail shown by the object. There are some cases we have dealt with, such as radar and sonar, where detection rather than imaging is the primary objective and the rule still applies; a radar wavelength of 10 m will detect a ship but would be useless for detecting an object a few centimetres in dimension. Again, if bats are to detect flying insects then the frequencies of the sound in the pulses they emit have to be high enough to give wavelengths similar to the dimension of the insect; a sound frequency of 50 kHz corresponds to a wavelength about 7 mm.

Now we are going to consider the problem of 'imaging' fundamental particles. There can be no question of producing a conventional image of such entities; a proton has a diameter of 10^{-15} m and an electromagnetic wave of such a wavelength has a frequency of 3×10^{23} Hz. We shall see that particle-physics theory indicates that the proton is not a fundamental particle but is a composite of other particles — *quarks.* Various combinations of quarks reveal themselves as other particles, none of which can be imaged as stationary entities. However, what can be done is to reveal their tracks through various kinds of matter and we must accept this as 'imaging' for these particles.

There are very many ways of detecting particle tracks and some of the most modern and effective do not show the tracks directly but rather give data from which the tracks can be deduced. However, there are a number of methods that give a direct visual representation of particle tracks and we shall just consider these, albeit they have mostly been superseded by other techniques.

Before describing ways of imaging particle tracks we first consider the nature of what is being imaged.

16.1. The Structure of an Atom

By the mid-1930s the basic structure of atoms was well understood. Almost all the mass is concentrated in a positively charged tiny nucleus, a few times 10^{-15} m in diameter. The nucleus consists of tightly packed protons, each with a unit positive charge, and uncharged neutrons, each of mass approximately equal to that of the proton. Surrounding the nucleus and occupying a space a few times 10^{-10} m in diameter are electrons, each with a unit negative charge and, for an overall neutral non-ionized atom, equal in number to the protons in the nucleus. The electrons, each with $1/1,836$ of the proton mass, contribute little to the mass but virtually all the volume of an atom.

The electrons occupy a sequence of energy levels in shells and the chemical properties of an element depend on the occupancy of the outermost shell, containing the most energetic and least well-bound electrons. The maximum number of electrons a shell can accommodate is governed by quantum-mechanical rules and elements with a filled outermost shell, such as the inert gasses helium, neon and argon, are chemically inactive. When atoms interact with each other to form compounds the main condition being satisfied is that each atom establishes an outer closed shell, a stable configuration

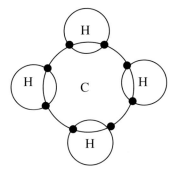

Figure 16.1   The molecule of methane held together by covalent bonds.

it attains either by it donating or receiving electrons to give ionic bonds (§15.6) or by sharing electrons to give *covalent bonding.* An example of covalent bonding, the methane molecule (CH_4), is shown in Fig. 16.1. The outer — and only — shell of an individual hydrogen atom contains one electron while that of carbon contains four. Sharing electrons gives a full outer shell, containing two electrons, for each hydrogen atom and a full outer shell of eight electrons for carbon.

The chemical properties of an element depend on the number of electrons, and hence protons it contains. It is more common than not that an element has several stable *isotopes* containing different numbers of neutrons in the nucleus. Thus a stable carbon atom can have either six or seven neutrons in its nucleus, the isotopes being represented symbolically as $^{12}_{6}C$ and $^{13}_{6}C$, respectively. The C is the chemical symbol for carbon, the 6 gives the number of protons and is the *atomic number* of carbon and the 12 and 13, the combined number of protons plus neutrons in the nucleus, are the respective *atomic masses* of the isotopes.

Some elements, or some isotopes of normally stable elements, are radioactive and give off emanations, which can be of three kinds. The first of these is α-particles, which are helium nuclei consisting of two protons and two neutrons. The fact that they are so commonly emitted indicates that stable groupings forming α-particles exist within the nucleus. These have only a short range in air and are completely absorbed by a sheet of paper. The second kind is β-particles, most of which are very energetic electrons but, in some

Figure 16.2   Tracks of β emission and a recoiling nucleus.

radioactive decays, are energetic *positrons*, particles with the same mass as that of the electron, but of opposite charge. The mechanism by which a nucleus emits a negative β-particle is that a neutron it contains decays into a proton and electron. Isolated neutrons are unstable with a half-life of about 15 minutes and it is only when they are bound into stable nuclei that they are stable. The emission of a positron is due to transformation of a proton into a neutron and a positron. This process requires an input of energy and so positron emission is rarer than the emission of electrons.

When observations were made of β-decay under conditions where the motions of both the nucleus and β-particle could be observed the tracks were seen as shown in Fig. 16.2.

It is clear from the geometry of this process that, just with the nucleus and β-particle being involved, momentum could not be conserved and it was postulated that a third, but invisible, particle, the *neutrino*, must also be involved. It had to be a particle with no charge and little, if any, mass. Its existence was predicted by the Austrian theoretical physicist Wolfgang Pauli (1900–1958; Nobel Prize for Physics, 1945) but only detected by an ingenious experiment carried out in 1959 by two American physicists, Clyde I Cowan (1919–1974) and Frederick Reines (1918–1998; Nobel Prize for Physics, 1995).

The third kind of emanation from radioactive decay is γ-radiation — very high-energy electromagnetic radiation. The wavelength of such radiation is shorter than 0.1 Å, sometimes much shorter, and that from radioactive decays can typically pass through 20 cm of lead.

The discovery of the positron led to a new concept in particle physics — the idea of an *antiparticle*. If an electron and positron come together they annihilate each other with the production of γ-radiation. Even more interesting is the fact that under suitable circumstances a γ-photon can be transformed into an electron plus a positron. This creation of matter from energy needs to satisfy Einstein's well-known equation, $E = mc^2$, where the energy of the photon of frequency ν is $E = h\nu$ and the mass being created is $2m_e$, where m_e is the mass of the electron. For electron-positron *pair production* the frequency of the γ-photon must satisfy

$$h\nu > 2m_e c^2. \tag{16.1}$$

Pair production requires an atomic nucleus to be in the vicinity to satisfy momentum conservation requirements. Photon energy over and above that required in creating the pair of particles, 1.022 MeV, will give them, and the atomic nucleus, kinetic energy. Showers of electrons and positrons are observed in cosmic radiation and are produced by pair production when high-energy γ-radiation interacts with atomic nuclei in the upper atmosphere.

When the behaviour of atomic electrons is analysed by quantum mechanics it is found that the states of all of them correspond to having an angular momentum that is an integral number of times a quantity $\hbar = h/2\pi$, where h is Planck's constant. However there is experimental and theoretical evidence that an electron has an intrinsic *spin angular momentum* with magnitude $\frac{1}{2}\hbar$ — it is said to have *half-integral spin*. There are other particles, such as the proton, neutron and neutrino, which also have half-integral spin. Since spin is a conserved quantity in any reaction then, apart from its role in momentum conservation, the neutrino is necessary to conserve spin when a neutron decays; a neutron, a half-integral spin particle, cannot decay into an even number of other half-integral spin particles — a third, the neutrino, is necessary to conserve spin. All particles with half-integral spin are known as *fermions* and those with integral spin — for example, α-particles are known as *bosons*. The two types of particle differ in the statistics of their energy distributions.

This summarizes knowledge about particles at the beginning of the 1930s, gained by a combination of small-scale bench-top experiments and the theoretical interpretation of those experiments. However, the subject of particle physics was about to embark on a much larger scale — and much more expensive — future.

16.2. Atom-smashing Machines

In 1919 the New Zealand physicist Ernest Rutherford (1871–1937; Nobel Prize for Chemistry, 1908) carried out a simple experiment in which, for the first time, a nuclear reaction was observed in which one type of atom was transformed to another. The reaction can be written as

$$\text{nitrogen} + \alpha\text{-particle} \rightarrow \text{oxygen} + \text{hydrogen}$$

or, symbolically, $\qquad {}^{14}_{7}\text{N} + {}^{4}_{2}\text{He} \rightarrow {}^{17}_{8}\text{O} + {}^{1}_{1}\text{H}.$ $\qquad$ (16.2)

The source of the α-particles, with energies several MeV, was a radioactive element; the energy of the α-particles had been sufficient to break up nitrogen atoms and the products of the collision had reformed into the nuclei seen on the right-hand side — an isotope of oxygen and a proton.

In 1932 the British physicist John Cockcroft (1897–1967) and his Irish colleague, Ernest Walton (1903–1995), both winners of the 1951 Nobel Prize for Physics, built an apparatus in which protons could be accelerated through a linear path up to energies of about 700 keV to bombard target materials with protons. With a lithium target they produced the reaction

$${}^{7}_{3}\text{Li} + {}^{1}_{1}\text{p} \rightarrow 2\,{}^{4}_{2}\text{He} \qquad (16.3)$$

In this reaction a proton (a hydrogen nucleus represented by ${}^{1}_{1}\text{p}$, which is the same as ${}^{1}_{1}\text{H}$), collides with a lithium nucleus to produce two helium nuclei (α-particles).

The total potential difference that could be generated to operate the Cockcroft–Walton linear accelerator was limited. Independently of Cockroft and Walton, the American physicist Ernest Lawrence (1901–1958; Nobel Prize for Physics, 1939) developed the *cyclotron*, a device in which, by a combination of electric and magnetic fields, charged particles travelled on an outward moving spiral path increasing their energy steadily as they did so. The principle involved can be understood by reference to Fig. 16.3. A bunch of charged particles is inserted at the centre of the cyclotron and travels in a curved path due to a magnetic field in a direction perpendicular to the hollow

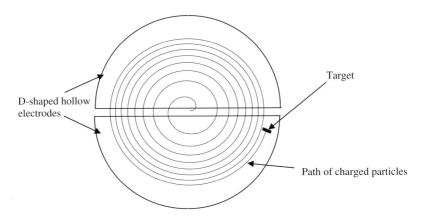

Figure 16.3 The cyclotron.

D-shaped electrodes. The period for one revolution of the particles is independent of the radius of the path and the radius increases with energy. The field between the two electrodes is periodically switched so that every time the particles cross from one to the other they are accelerated. In this way the energy of the particles steadily increases without the need for a very high potential difference across the plates. Eventually they strike a target to generate new products. If the particles become relativistic, i.e. their masses appreciably changed because of relativistic effects, then their period of revolution will depend on their energy and then it is necessary to change the timing of the potential differences across the electrodes to maintain the constant acceleration; such a device is called a *synchrocyclotron*, which can be used to produce particles with energies of up to 700 MeV.

The next development in producing atom-smashing devices is characterized by the Stanford linear accelerator, in the form of an evacuated pipe of length 3.2 km. The particles, in tight bunches, move through a series of separated copper tubes, as shown in Fig. 16.4; each time they reach the gap a potential difference is present between the tubes on the two sides of the gap that accelerates the particles. The potential difference needs to be present only when the particles are traversing the gap and is switched from one gap to the next as they travel along the tube. In this way an effective

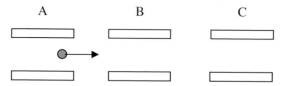

Figure 16.4   A basic component of a linear accelerator. A, B and C are copper tubes and the grey circle a bunch of charged particles.

accelerating potential difference of $30\,\mathrm{GV}$[1] is obtained with only a source of comparatively small potential difference being required.

The final type of particle accelerator is essentially a synchrotron, as described in §15.4, except that the particles need not be electrons and the accelerator is optimized to produce high energy particles rather than radiation. The most powerful machine of this kind is the *Large Hadron Collider* (LHC) run by CERN (Centre European de Recherche Nucléaire), 27 km in circumference and situated on the Swiss–French border. Its 9,300 super-conducting bending magnets guide the particles round the circuit and it is possible to accelerate counter-rotating beams of protons or heavy particles in separated paths up to speeds of 99.999 999% of the speed of light and then deflect them so that they smash into each other.

Modern atom-smashing equipment is capable of producing many exotic new particles, which will now be described.

16.3. Many More Particles

The stability of an atomic nucleus is, *prima facie*, rather unexpected. It would seem that normal repulsive electrostatic forces between the positively charged protons should cause the nucleus to fly apart. Clearly there is some other force in play and the Japanese physicist Hideki Yukawa (1907–1981; Nobel Prize for Physics, 1947) explained the nature of that force in 1935. His explanation depended on the idea of *exchange forces*, mediated by an exchange particle passing to and fro between the particles experiencing the force. An analogy for a particle producing a repulsive force is to imagine two people standing

---

[1] $1\,\mathrm{GV} = 10^9\,\mathrm{V}$.

on a frictionless surface passing a ball backwards and forwards. Every time one of them throws the ball to the other he experiences a backwards force, as he also does when he catches the ball thrown by the other person. The two steadily move apart as though there was a repulsive force operating between them. A similar analogy for an attractive force is more difficult to imagine in the same terms since it requires the ball to have a negative mass.

Yukawa postulated that when nuclear particles are very close they experience an attractive force due to an exchange particle, the properties of which he could deduce theoretically as that it was a boson with mass about 200 times that of the electron. This particle he called a *meson*.

Before high-energy atom-smashing machines became available, experiments were performed that carried photographic plates to a high altitude using balloons. Tracks on the plates showed the passage of particles produced by the collisions of high-energy cosmic ray particles with atoms within the plate emulsion. A pioneer in this work was the British physicist Cecil Powell (1903–1969; Nobel Prize for Physics, 1950) who discovered a particle with mass 207 electron masses that he thought was the meson, but it turned out to be a fermion, a different particle given the name *muon*. The use of cosmic rays to provide particles for collisions was a rather hit-and-miss technique and with the advent of the atom-smashing machines many new and exotic particles, with different masses, charges and other properties, have been discovered, including Yukawa's meson.

We have already met the positron, the antiparticle of the electron, and collision experiments have shown that to every particle there is an antiparticle, for example, there are antiprotons, with a negative charge and neutral antineutrons and antineutrinos. By convention the particle produced by the break-up of a neutron into a proton and electron is called an *electron antineutrino* whereas that produced by break-up of a proton is known as an *electron neutrino*. Although they are both neutral particles, a meeting between a neutrino and an antineutrino would lead to their mutual annihilation, with the production of a γ-ray photon.

From experiments, a reasonably small number of different high-energy particles can be identified. There is a group of six particles, called *leptons* and related to the electron, which are fundamental in the sense that they cannot be decomposed into other particles. They are all fermions with a unit negative charge, but vary greatly in their masses. They are shown in tabular form in Fig. 16.5 and there are also six corresponding antiparticles. The muon is the particle discovered by Powell.

1 electron	**207** muon	**3477** tau
electron neutrino	muon neutrino	tau neutrino

Figure 16.5   The six leptons with masses in electron units.

There are other kinds of exotic particles — for example, *omega*, *lamda* and *kaon* particles — that, together with protons and neutrons, can be decomposed into basic components. In 1964 two American theoretical physicists, Murray Gell-Mann (b. 1929; Nobel Prize for Physics, 1969) and George Zweig (b. 1937), independently proposed a model for these components, which are now known as *quarks*.

There are six types of quark, with different characteristics described as their *flavours* — *up* (u), *down* (d), *strange* (s), *charm* (c), *bottom* (b) and *top* (t) — all fermions with spin $\frac{1}{2}$. Their charges are all multiples of $\frac{1}{3}$ of an electron charge and for the six flavours they are

$$u\left(\frac{2}{3}\right) \quad d\left(-\frac{1}{3}\right) \quad s\left(-\frac{1}{3}\right) \quad c\left(\frac{2}{3}\right) \quad t\left(\frac{2}{3}\right) \quad b\left(-\frac{1}{3}\right)$$

Associated with each quark there is an *antiquark* with opposite charges, represented by

$$\bar{u}\left(-\frac{2}{3}\right) \quad \bar{d}\left(\frac{1}{3}\right) \quad \bar{s}\left(\frac{1}{3}\right) \quad \bar{c}\left(-\frac{2}{3}\right) \quad \bar{t}\left(-\frac{2}{3}\right) \quad \bar{b}\left(\frac{1}{3}\right)$$

Combinations of three components of the up-down type give rise to protons, neutrons and their antiparticles; since each quark is a fermion then combinations of odd numbers of them will also be fermions.

$$\text{Proton} \quad u + u + d \quad \text{has charge} \quad \frac{2}{3} + \frac{2}{3} - \frac{1}{3} = 1$$

$$\text{Antiproton} \quad \bar{u} + \bar{u} + \bar{d} \quad \text{has charge} \quad -\frac{2}{3} - \frac{2}{3} + \frac{1}{3} = -1$$

$$\text{Neutron} \quad d + d + u \quad \text{has charge} \quad -\frac{1}{3} - \frac{1}{3} + \frac{2}{3} = 0$$

$$\text{Antineutron} \quad \bar{d} + \bar{d} + \bar{u} \quad \text{has charge} \quad \frac{1}{3} + \frac{1}{3} - \frac{2}{3} = 0$$

Particles formed by combinations of three quarks, such as the proton and neutron, are known as *baryons* and other baryons are *lamda particles*. The three types of lamda particle, and their compositions are

$$\Lambda^0 \quad u + d + s \quad \text{with charge} \quad \frac{2}{3} - \frac{1}{3} - \frac{1}{3} = 0$$

$$\Lambda_s^+ \quad u + d + c \quad \text{with charge} \quad \frac{2}{3} - \frac{1}{3} + \frac{2}{3} = 1$$

$$\text{and} \quad \Lambda_b^0 \quad u + d + b \quad \text{with charge} \quad \frac{2}{3} - \frac{1}{3} - \frac{1}{3} = 0$$

with the corresponding antiparticles.

Quarks can also come together in pairs involving one quark and one antiquark to give various kinds of meson, which are bosons. As an example of this, *kaons*, also known as K-mesons, are formed from pairs of quarks one of which must be the strange quark, or its antiquark, and the other either an up or down quark or antiquark. Although there are some claims to the contrary it is doubtful that quarks themselves have ever been observed.

16.4. Direct Imaging of Particle Tracks

The term *particle detector*, as used in particle physics, may have one of two meanings, either a device that simply counts and records particles or one that actually follows the path of a particle and displays it in some way. The display techniques are also of two forms. In the first there is a straightforward image of the track that can be directly viewed or photographed. These ways of forming an image are dependent on the chemical or physical action of the particles on some material — a photographic plate or a fluid — that changes its appearance. Since these processes directly give images they are of the greatest interest in the context of this book. However, they suffer from the disadvantage that the rate at which they can collect information is limited. For this reason methods based on electronics are now preferred for recording the tracks of high-energy particles. These latter methods, which include *wire chambers* and *silicon detectors*, give electrical signals that can be interpreted by computer calculations into three-dimensional tracks, but these will not be further considered since we have more direct ways of producing images.

16.4.1. *Photographic plates*

The first indication that elementary particles could be detected by photography came from the work of the French physicist, Henri Becquerel (1852–1908; Nobel Prize for Physics, 1903). Becquerel was an expert on the topic of phosphorescence, the process by which a substance absorbs and stores energy by irradiation and then, over time, emits electromagnetic radiation at some characteristic frequencies. In 1896 he heard a lecture about Röntgen's work on x-rays and incorrectly concluded that x-rays were some form of phosphorescent emission. He tried experiments with various phosphorescent materials, exposing them to sunlight and then placing them on top of a photographic plate wrapped in black paper. If the materials emitted x-rays then the plate would be blackened. He found by accident that a salt containing uranium would blacken the plate even if it had not been initially irradiated. Later it was ascertained that particles

Figure 16.6   A high-energy cosmic-ray particle striking an atom within a photographic plate (NASA).

produced by the radioactive decay of uranium was the cause of the blackening.

Cecil Powell was the first to develop systematically the technique of using stacks of photographic plates to detect the tracks of high-energy particles and this method is still occasionally used, although it has been largely displaced by other methods. An example of the kinds of tracks that can be observed is shown in Fig. 16.6, which shows the outcome of the collision of a high-energy cosmic-ray particle with an atom in the photographic plate. The cosmic-ray particle comes in from the right and produces a myriad of other particles leaving narrower tracks.

16.4.2. *The Wilson cloud chamber*

In 1911 the Scottish physicist Charles Wilson (1869–1959; Nobel Prize for Physics, 1927) invented a device, know as a *cloud chamber*, for directly viewing the tracks of particles coming from radioactive materials. The basis of the cloud chamber is the production of a region saturated with water vapour; when a particle moves through the region it reveals its track by leaving a trail of water droplets.

A very simple design of cloud chamber is illustrated in Fig. 16.7. The space above the fluid, which was water in Wilson's original design but is now usually methanol (CH_3OH) or ethanol (C_2H_5OH), is saturated with vapour. When the piston is moved downwards then, due to expansion, the saturated air cools thus reducing the amount of vapour required to achieve saturation. The air becomes supersaturated and a mist forms as tiny droplets begin to condense.

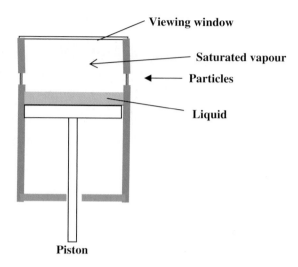

Figure 16.7   A Wilson cloud chamber.

During this period a high-energy particle passing through the region will leave a trail of larger droplets, which may be viewed or photographed through the glass window. To facilitate viewing the top of the piston is painted black.

It is possible to have the particle source inside the cloud chamber if it is a radioactive source, otherwise the particles move inwards through a thin part of the cloud-chamber wall. The sign of the charge on particles, or whether they have no charge, can be ascertained by the addition of a magnetic field perpendicular to the motion of the particles and noting either their direction of deflection, which gives the sign of the charge, or that they are not deflected, which shows that they are uncharged.

In 1928 the British physicist Paul Dirac (1902–1984; Nobel Prize for Physics, 1933) developed a theory for fermions that predicted the existence of the positron. The American physicist Carl Anderson (1905–1991; Nobel Prize for Physics, 1936) confirmed this prediction in 1932. Anderson was using a Wilson cloud chamber across which was a high magnetic field to determine the types, charges and energies of cosmic ray particles. Some of the tracks were characteristic of electrons but seemed to bend in an opposite direction

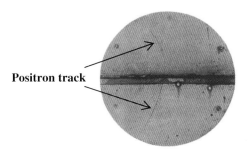

Figure 16.8   Carl Anderson's image of a cloud-chamber positron track.

in the magnetic field, indicating that they had a positive charge. However, the direction of bending depends not only on the sign of charge but also on the direction of motion of particles. To confirm that what he was seeing were positively charged Anderson placed a lead plate within the cloud chamber. After passing through the lead plate the particles had less energy and would bend more sharply. Figure 16.8 shows the image of the cloud chamber track published by Anderson that unambiguously confirmed that what was being seen was a positron. The curvature is greater above the lead plate showing that its passage was upwards.

16.4.3. *The bubble chamber*

In 1954 the American physicist, Donald Glaser (b. 1926; Nobel Prize for Physics, 1960) invented a new kind of particle-track detector. The basis of this technique is to create a metastable superheated liquid — that is a liquid that is not boiling but is above its boiling point. When such a liquid is disturbed, for example by an injection of energy due to the deceleration of an elementary particle, then local boiling takes place as revealed by a trail of tiny bubbles in the liquid.

Figure 16.9 shows a schematic illustration of a bubble chamber. The liquid normally used is hydrogen that, at atmospheric pressure, boils at 20.46 K. The boiling point depends on the pressure, falling with decreasing pressure; it is well known that on high mountains it is impossible to make a really hot drink because water boils at well below its boiling point at sea level. The hydrogen is initially

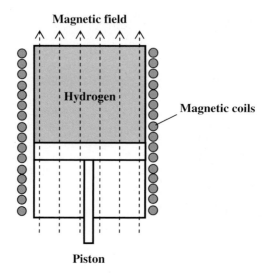

Figure 16.9   A bubble chamber.

kept at a pressure of 5 atmospheres and at just below the boiling point for that pressure. Then the piston is withdrawn, quickly reducing the pressure to 2 atmospheres, at which the boiling point is below the prevailing temperature. The liquid is now superheated and particles passing through it leave a trail of bubbles. When the bubbles have grown to a size at which they are easily visible, about 1 mm, flash photographs are taken from a number of viewpoints, the analysis of which enables the tracks to be found in three dimensions.

The first bubble chambers were quite small, up to 30 cm or so across, but recent ones are up to 4 m in diameter. A larger chamber makes the measurement of the curvature of tracks much more accurate. A beautiful example of a bubble-chamber image, produced by CERN and shown in Fig. 16.10, contains a number of tracks of different kinds of particle.

Many new particles have been discovered using bubble chambers. Figure 16.11 shows the first image of the result of a neutrino collision with a proton in hydrogen liquid; the hydrogen is not only the medium in which tracks are revealed but also provides proton targets

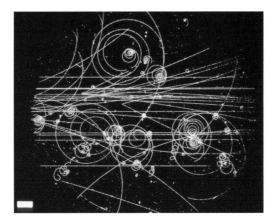

Figure 16.10   Bubble-chamber tracks showing many particles (CERN).

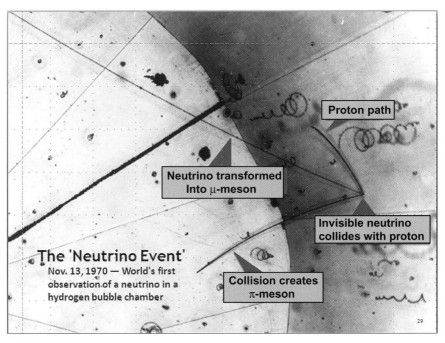

Figure 16.11   The first bubble-chamber image of the result of a neutrino collision with a proton.

for the oncoming radiation. In the annotated image, produced in 1970, the invisible neutrino comes in from the right, strikes a proton and then produces the tracks of three particles — the proton, a μ-meson and a π-meson.

Bubble chambers give excellent images of particles and collision events but their data acquisition rate is rather slow, which is why electronics-based detectors have largely superseded them.

Index